# ATLAS OF THE NATURAL WORLD

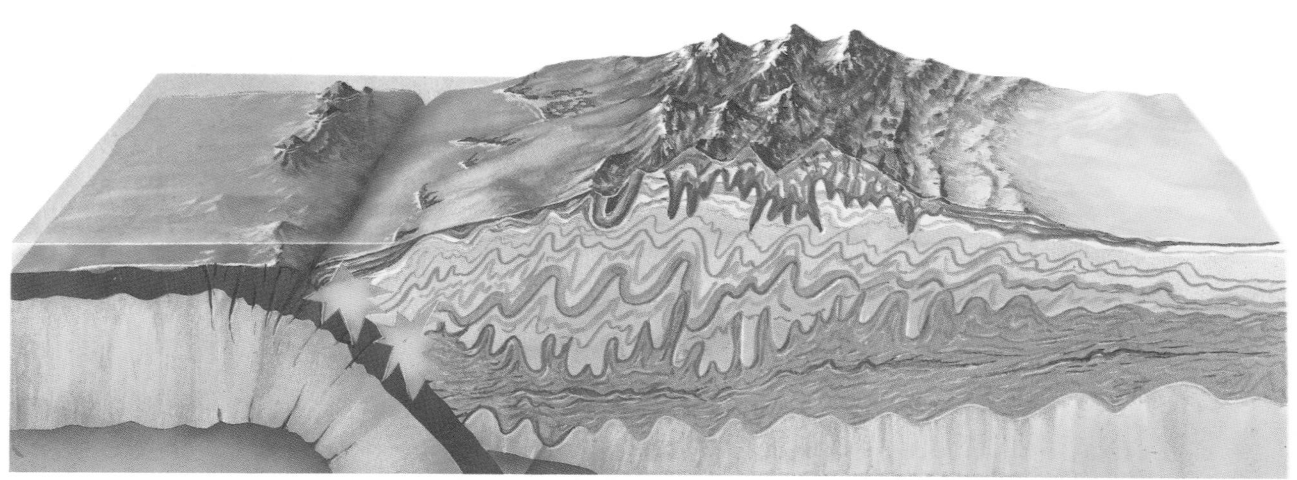

# ATLAS OF THE NATURAL WORLD

ROBERT MUIR WOOD

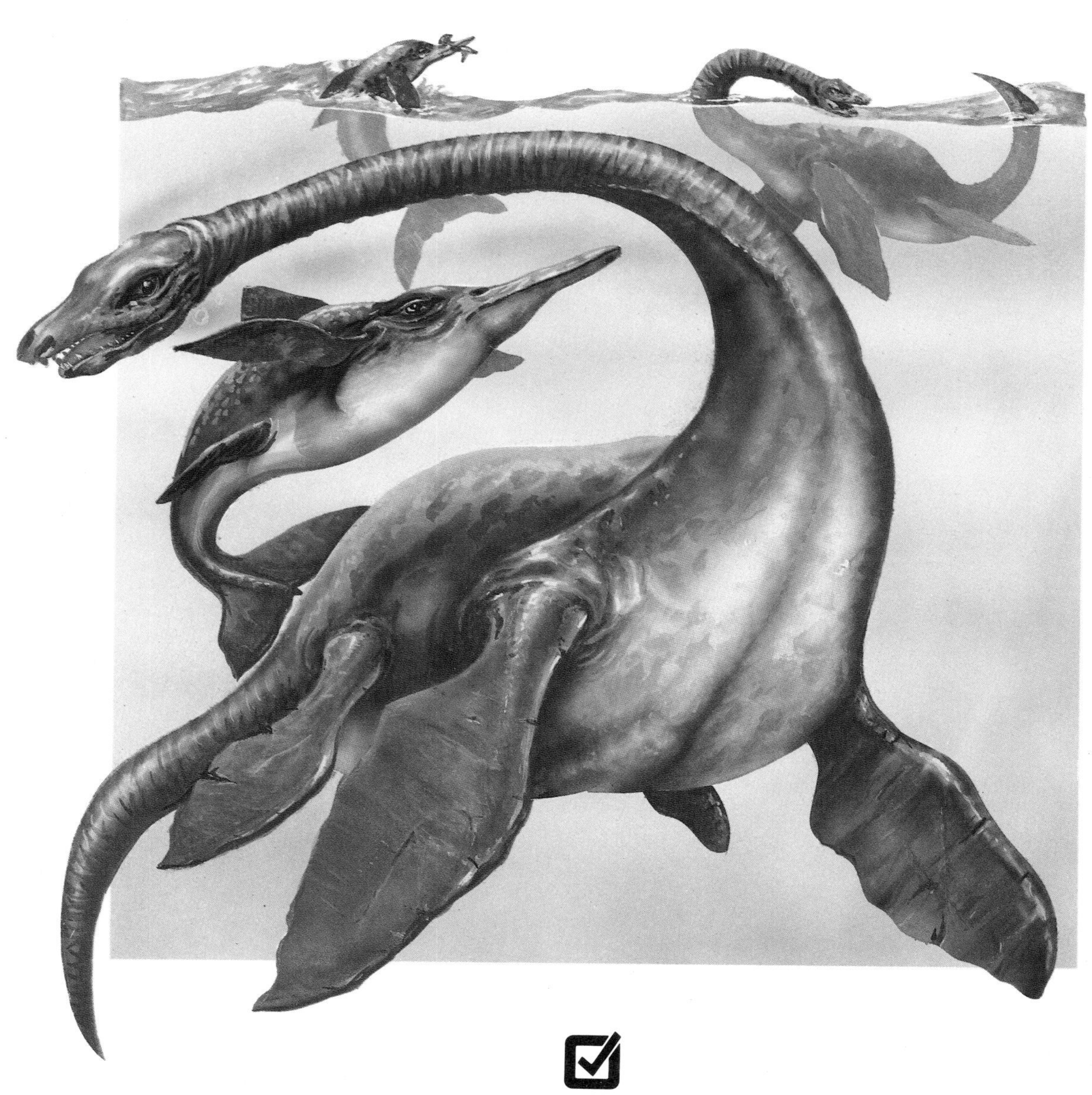

Facts On File
New York • Oxford

For information contact:

Facts On File, Inc.
460 Park Avenue South
New York NY 10016

Library of Congress Cataloging-in-Publication Data

**Muir Wood, Robert**
Atlas of the natural world.
Includes index
Summary: Text and maps depict and explain the natural world and the structure and processes of the Earth.
1. Natural history—Maps. 2. Physical geography—Maps. 3. Natural history. 4. Physical geography. 5. Ecology. [1. Natural history. 2. Natural history—Maps. 3. Physical geography. 4. Physical geography—Maps.] I. Title.

G1046.C1M5 1990 508'.0022'3—dc20 89–675217
ISBN 0–8160–2131–7

Facts On File books are available at special discounts when purchased in bulk quantities for businesses, associations, institutions or sales promotions. Please call our Special Sales Department in New York at 212/683–2244 (dial 800/322–8755 except in NY, AK or HI).

**An Ilex Book**

Created and produced by Ilex Publishers Limited
29–31 George Street, Oxford OX1 2AJ, England

Designed by Phil Jacobs
Illustrated by Mike Saunders
Edited by Nicholas Harris
Maps by Zoë Goodwin and Janos Marffy
Cover artwork by Steve Weston

Typesetting by Opus, Oxford
Colour separations by Scantrans Pte. Ltd
Printed in Spain

10 9 8 7 6 5 4 3 2 1

This book is printed on acid-free paper.

# CONTENTS

# The origin of the Earth

The story of how the Earth began is one we have to reconstruct from small clues, for almost all the evidence was destroyed in the very act of creation. Something we can, however, be sure of is, most importantly, the Earth's date of birth. Careful measurements of the concentration of the daughter products of radioactive decay confirm that the Earth and all the other planets and meteorites in our solar system were born some 4.6 billion years ago. We know also where the elements that form the Earth originated. Like everything else in the solar system, the rocks of the Alps, the rings of Saturn, the water in the oceans, the bark of a tree, and the retina of the human eye – are all made out of star dust.

We can also hazard some good guesses as to how the Earth originally formed. In a cloud of dust and gas, the remains of some former star, clumps of denser material began to collide with one another, eventually becoming the planets of the solar system.

A cloud of dust and gas becomes a whirling disk nine billion miles wide. The core evolves into the sun while particles of rock or ice collide, stick together, and begin to grow into planetary bodies.

### The planets form

The Earth was the third mass to develop in orbit around the Sun, and originally possessed an enormous atmosphere of hydrogen and helium that weighed more than the rocky material beneath. Yet probably because the young Sun was particularly massive and vigorous, the original atmosphere was blasted off all th inner planets — Mercury, Venus, Earth, and Mars — by a vicious solar wind. Thes planets must have looked like gigantic comets, with thick atmospheric tails streaming away from the Sun. Only the planets further from the Sun — Jupiter, Saturn and Uranus — have preserved thei original atmospheres. Today, while the rocky centers of these outer planets are scarcely bigger than the Earth, they are shrouded beneath atmospheres, oceans and ice caps of hydrogen and helium, tens of thousands of miles deep.

## The second atmosphere

As the Earth accumulated, its interior started to melt, both from the heat released by the radioactive decay of some of its atoms, and from the inward collapse of materials. Melting led to volcanoes, which were almost universal over the face of the Earth. The Moon was very close to the Earth at this time, orbiting as fast as every six and a half hours. Blazing with intense ultraviolet radiation, the Sun scurried from horizon to horizon in a few hours.

After the Sun had quieted into a calmer central source of heat and light, gases such as water, carbon dioxide, nitrogen and hydrogen chloride, originally trapped inside minerals, emerged from the Earth in volcanic eruptions and slowly formed a new atmosphere. At first this atmosphere was hot and thin, but as it thickened, and cooled, so one night, away from the Sun's fierce heat, it rained.

These first rains were extremely acidic and reacted with the rocks to form basins of salt where the water evaporated. In time the ponds that formed in topographic depressions joined to form seas, and expanded until every land-bridge was drowned and there was a world-encircling ocean, subject to gigantic lunar tides.

These oceans saved the Earth from Venus's fate, where carbon dioxide liberated from the interior created a "greenhouse effect." Visible light from the Sun is reradiated from the surface of Venus in the infrared spectrum and as carbon dioxide is opaque to infrared, this energy is trapped. The surface temperature on Venus today is 880°F (472°C): hot enough to melt lead.

## Before the Earth was born

Soon after midnight on February 8, 1969, a large meteorite plunged out of the sky near Pueblito de Allende in Mexico. American scientists quickly realized this was a massive specimen of one of the rarest meteorite-rocks, a carbonaceous chondrite. The meteorite (christened Allende) was composed of the original particles that had crystallized out of an interstellar dust cloud before the Earth was born. From a study of Allende elements, scientists found traces of the gas xenon, the daughter product of the radioactive decay of a short-lived isotope of iodine. Such an isotope could not have been incorporated into Allende unless the star-furnace had been forging new elements as recently as 50 million years before the cosmic dust began to form the solar system.

# The mantle motor

After the Earth formed, its interior melted and the denser molten metals sank into the center of the planet to form a liquid metal core. Over 4.5 billion years, the liquid has crystallized into a solid metal center.

Above the core lies the Earth's mantle, making up 83 percent of the Earth's volume. The mantle is so hot that its rocks can flow like a sluggish liquid. The Earth loses heat only through its outer surface, so the top of the mantle is colder than its base. This creates convection currents. The hotter, less dense mantle rises, while the colder, denser mantle sinks.

As the mantle currents rise, the reduction in pressure causes the rocks partly to melt. A lighter "basaltic" magma bleeds off to the surface where it cools and hardens into a thick skin. This skin, largely hidden beneath the oceans, covers over 70 percent of the Earth's surface.

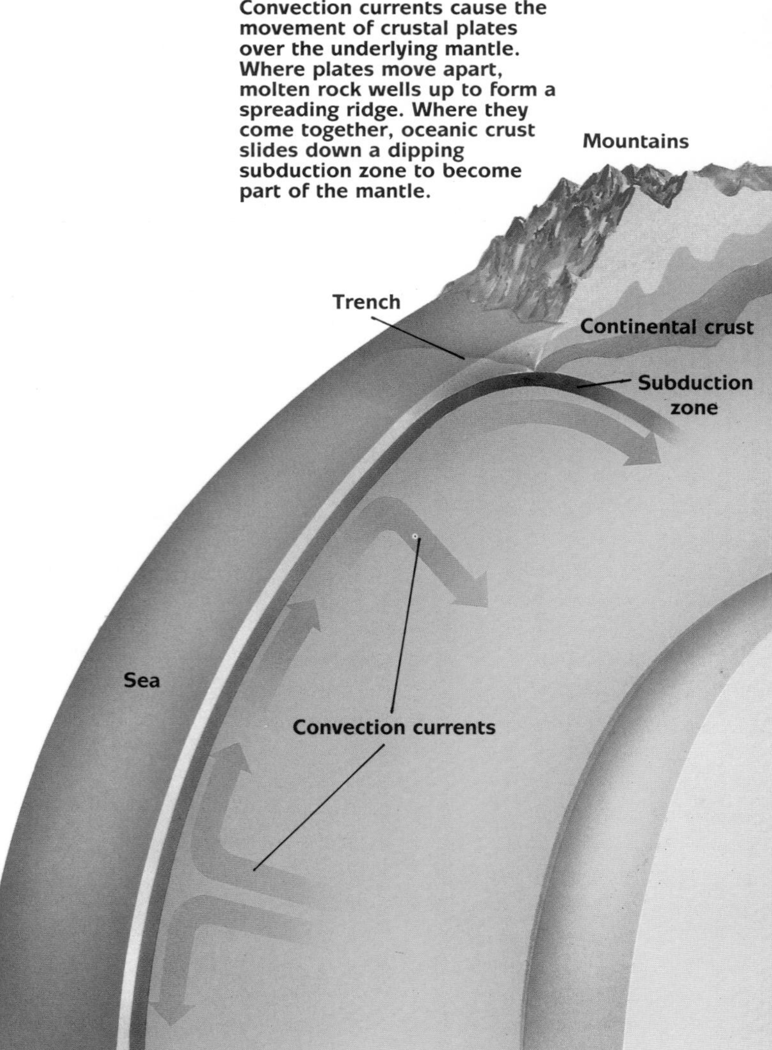


Convection currents cause the movement of crustal plates over the underlying mantle. Where plates move apart, molten rock wells up to form a spreading ridge. Where they come together, oceanic crust slides down a dipping subduction zone to become part of the mantle.

## Gaining access to the mantle

It is easy to think that the cold hard rocks we find at the surface are typical of the Earth as a whole. Volcanoes, where molten rocks from the mantle break through to the surface, provide more realistic "windows" into the Earth's interior. While the magmas that erupt from volcanoes do not have the same chemistry as the mantle (they are "fractionates" — chemicals that melt at the lowest temperatures), certain magmas carry with them blocks of mantle rock.

In the 1960s plans were made to drill through the 3-mile (10 km) thick crust below the oceans to take samples of the mantle beneath the Moho (or Mohorovičić Discontinuity), the boundary between the crust and the mantle. This Mohole Project grew too expensive and was abandoned. However, in many regions where continents have collided, sections of the mantle have become pinched up to emerge at the very surface. These rocks, known as peridotites, are found in the sites of many former mountain ranges, such as the Alps, Pyrenees, Urals, and the Cascade Mountains of northern California.

## Continental rafts

The rocks of the continents are lighter than the mantle below. These masses of less dense rock that make up the continental crust have accumulated over the past 4 billion years, and float like rafts, some 18 to 35 miles (29 to 56 km) thick, on the underlying mantle.

Because the ocean crust is the upper surface of the mantle convection, it is involved in the mantle circulation. As the convection currents move a few inches each year, the rate at which the ocean floor is replaced means that there is no ocean crust older than 200 million years.

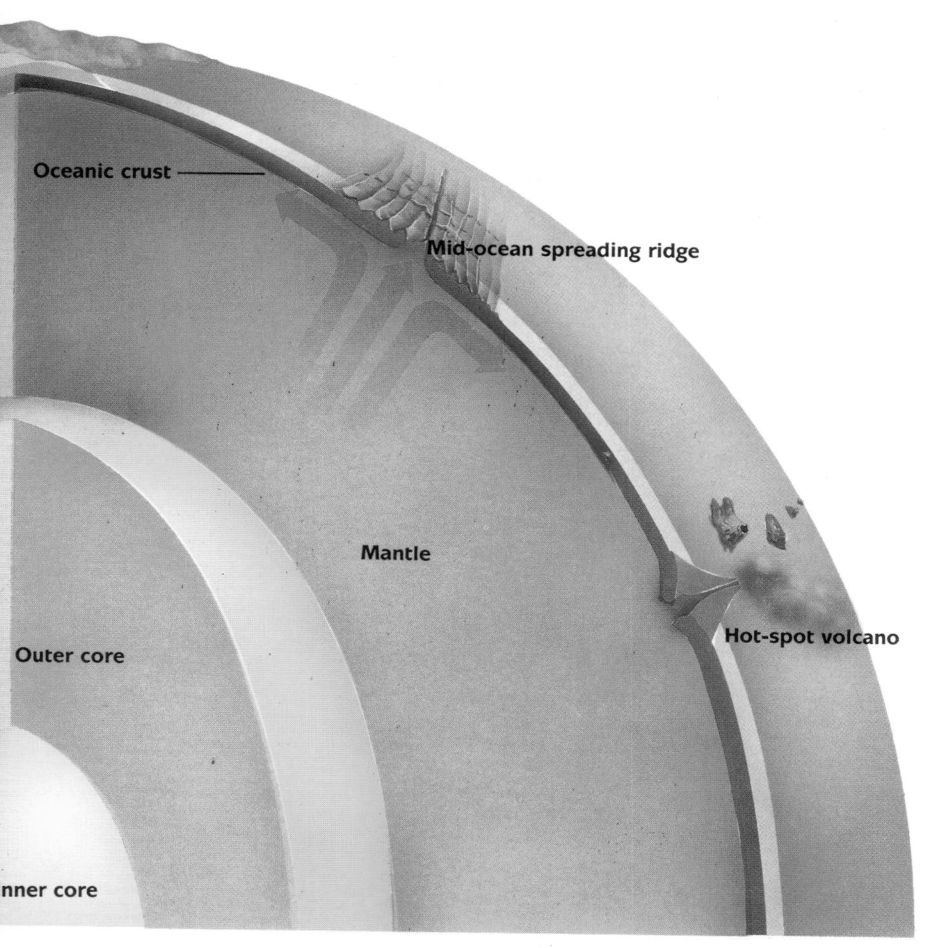


## Hot-spots

From deep down in the mantle, occasional plumes of very hot material rise up and burn holes through the mantle and crust emerging at the surface as large basaltic volcanoes. As a continental plate drifts over this plume, a line of volcanoes is scorched into the crust. The most famous example of such a "hot-spot" is the Hawaiian island chain, formed during the last 50 million years as the Pacific plate has drifted towards the west over a rising chimney of deep hot mantle. A major hot-spot also lies beneath Iceland. Sixty million years ago, this plume showered volcanoes over a wide region from southwest England to Eastern Greenland.

## The plates

While the continents drift, their motion is locked into the movement of the ocean crust and underlying mantle. The whole of the Earth's outer shell, down to depths of around 60 miles (96 km), moves as rigid "plates," driven by the mantle motor. The slow motion of the continental rafts can divide populations of animals or plants so that they evolve along distinct paths. Hundreds of millions of years later, these populations may once more be brought back into contact and competition with one another.

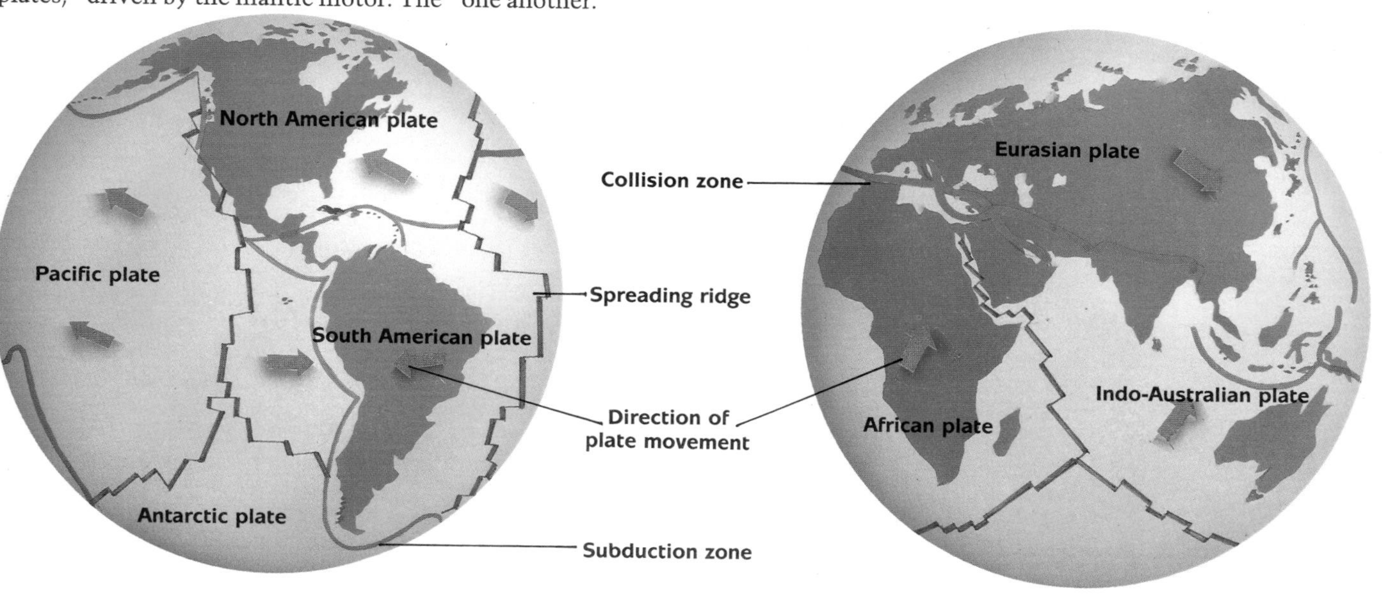

# The Earth's climate

Climate is the total weather over the course of several years, a combination of temperature and precipitation (rain and snow). It consists of not just the averages of heat or rainfall, but also the range of extremes such as cold winters and droughts. Plants are highly sensitive to climate, and the areas in which different plants flourish define the climatic zones.

Climate results from the solar powered heat-engine that drives the Earth's weather. The heat of the Sun is concentrated in the equatorial regions, while at high latitudes the Sun's energy is diffused over the surface and lost in glancing through the atmosphere.

The hot equatorial air, being less dense, rises and moves away from the Equator, sinking down to the surface somewhere between one-third and one-half of the way to the poles. At the North and South Poles the cold heavy air spreads slowly outwards into lower latitudes. It is the interaction between this polar air, and the sinking equatorial air, that produces the most variable weather patterns in the mid-latitudes.

The location of continental masses around the poles is particularly critical for global climate. If all the polar regions were oceanic then the Earth would be significantly warmer than if the polar regions were continental.

The climatic north-south contrast, caused by an Arctic *ocean* and an Antarctic *continent* means that French vineyards (right) are at the same latitude as a sea-level glacier in Chile (below).

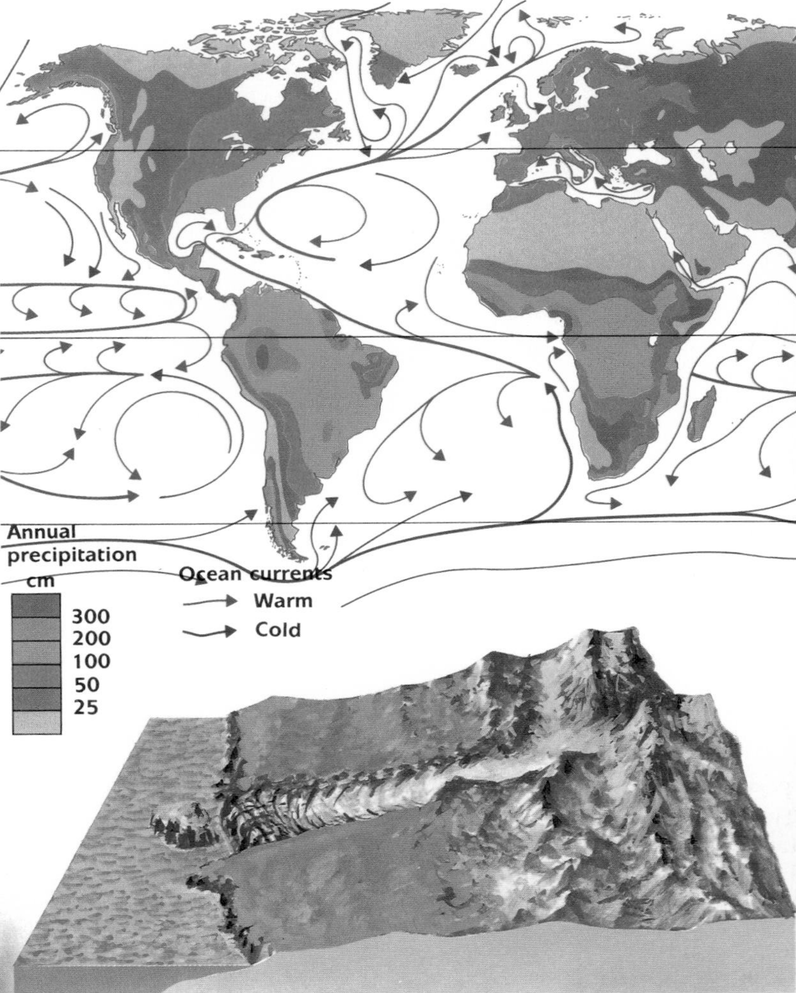


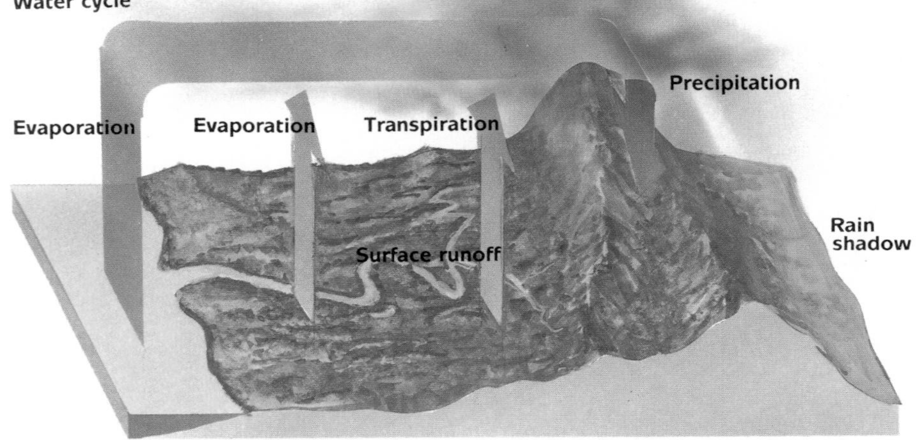


### Clouds and rain

The global movement of air masses controls the distribution of rain and snowfall. Water evaporates into the air, the water content increasing with warmer temperatures. Once this saturated air cools the water reappears and eventually falls to Earth. Cooling usually occurs when air rises, as over mountains, or in convective columns above a heated land surface. Highest rainfall is found over mountains receiving winds from nearby warm oceans. Lowest rainfalls occur in continental interiors where winds perpetually blow offshore, and in the shadow of mountains, which have already extracted the atmospheric moisture.

## The heat-engine

Warm equatorial air rises and moves towards the poles, while nearer the Earth's surface, cold polar air flows in the opposite direction. This pattern is complicated by the Coriolis Force, caused by the Earth's rotation.

As the Earth spins around its axis, every object at the Equator rotates at 1,020 mph (1,634 kph). Closer to the poles, at 30°N, this rotation slows to 900 mph (1,440 kph), while at the poles themselves the speed of rotation is zero.

In the oceans the Coriolis Force sets up the enormous slow circulations, the gyres, clockwise in the northern hemisphere and counterclockwise in the southern hemisphere.

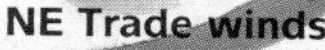


## The influence of oceanic currents

The oceans are full of enormous 'rivers' transferring water masses from one latitude to another. The flow of these rivers is layered according to density. The Gulf Stream carries hot, less dense water but larger currents run deeper down, where the water is barely above freezing. Over the oceans the water absorbs heat keeping summers cool, and subsequently releasing the heat to keep winters warm. Oceanic islands, even at relatively high latitudes, have equable climates: Iceland has warmer winters than New York, 1,500 miles (2,400 km) to the south. The Earth's climate is at present relatively cold because of the small enclosed frozen ocean at one pole and the large continent at the other.

## The effect of continents

Land absorbs heat much less readily than the sea. Because of this, seasonal changes in solar heating become amplified over the continents, where summers are hot and winters bitterly cold. These changes in continental temperatures have a major impact on climates. For example, hot air rising over Asia pulls in moist equatorial air from the Indian Ocean that brings torrential rain, called the "monsoon," to India. In the winter dry winds move outward from the cold Asian interior.

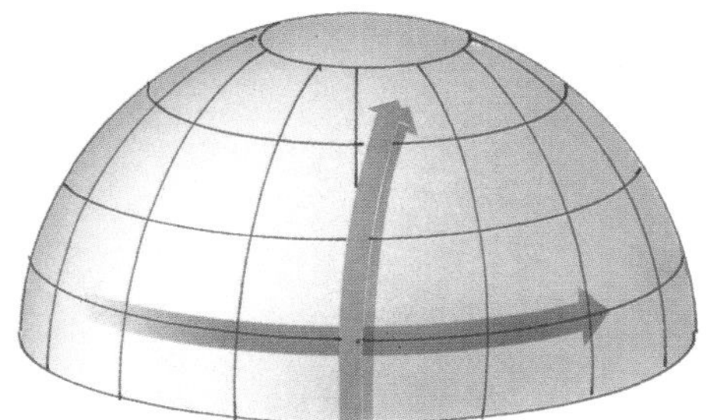

## Demonstrating Coriolis Force

If a bullet is fired due north from the Equator, as the ground moves more slowly, so the bullet will appear to start drifting eastwards. This phenomenon, Coriolis Force, must be taken into account by airline pilots (and gunners) intending to reach their targets.

# The origin of life

The greatest mystery in planet Earth's development concerns the origin of life. The odds of life happening have been calculated by the British astrophysicist, Fred Hoyle, as equivalent to the possibility that 'a tornado sweeping through a junkyard might assemble a Boeing 747 from the materials therein.' The designers of an airplane make a blueprint for the builders to follow. How could life have emerged without such a plan?

We can never hope to see the beginnings of life preserved in the rocks. There are no known sediments from the time when life first appeared on Earth, and we are dealing with processes so tiny that there would be no chance of them being preserved anyway. Also there is every indication that the creation of life, or at least the kind of life that now populates Earth, only happened once. This means that if we go back some 4 billion years, all living creatures – from trees to pond slime, from elephants to fleas – share a common ancestor.

**Building-blocks**
The simple building-blocks are unlikely to have combined without assistance. The walls of bubbles (far right), or the surfaces of natural minerals (right), have been suggested as places where molecules could have joined to form more complex chemical structures that could reproduce themselves.

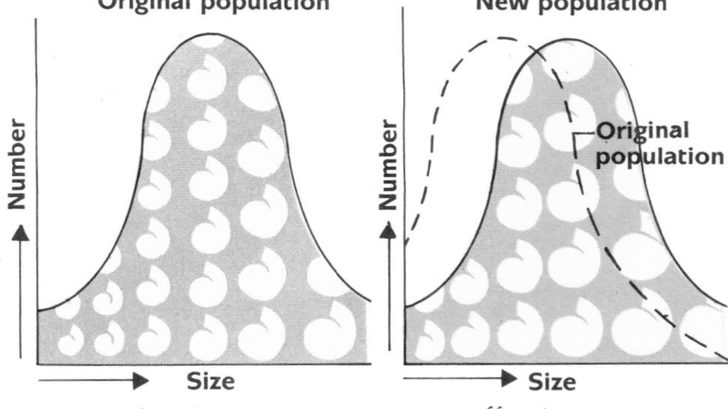


**Natural Selection**
Any population of organisms contains a range of individuals that vary slightly from one another in all aspects of shape, color, and behavior. The range of sizes of individuals typically shows a normal distribution. If large individuals produce more offspring – perhaps because food is plentiful – the average size of the population will, over many generations, gradually increase. Similar processes of natural selection have caused the evolution of all the variations among plants and animals.

**Evolution**
All living forms have the ability to pass on to the next generation a blueprint for life and growth, known as the genetic code. Through tiny variations of this code, caused when the genes from two parents are mixed or through natural mutation, organisms are created that differ from other members of the same species. If some members of a species are better adapted to their environment (possessing perhaps an improved method of surviving drought or of detecting prey) they may have more children. This process is called natural selection. Over many generations a successful, new adaptation can dominate a population – a process known as evolution. Organisms descended from the same parents may evolve differently if they become separated into smaller, isolated populations with different environmental circumstances.

**An extra-terrestrial origin?**
Some scientists find the idea of life originating on Earth so improbable that they believe it must have begun elsewhere in the Universe and arrived here in a spacecraft or in dust from outer space. Perhaps the whole history of life on Earth is the result of a remarkable experiment, performed by extremely intelligent creatures. These explanations, however, simply shift the place of origin elsewhere.

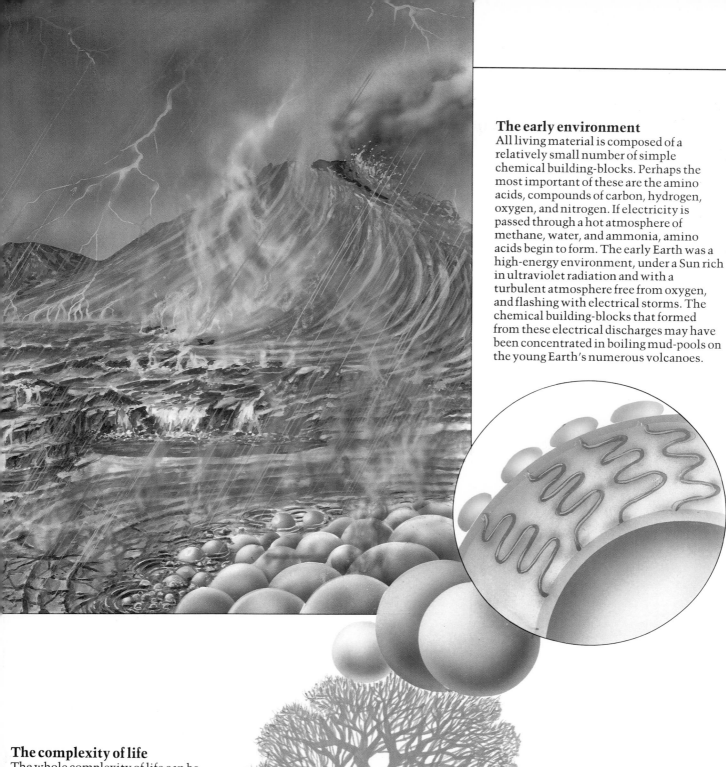

## The early environment

All living material is composed of a relatively small number of simple chemical building-blocks. Perhaps the most important of these are the amino acids, compounds of carbon, hydrogen, oxygen, and nitrogen. If electricity is passed through a hot atmosphere of methane, water, and ammonia, amino acids begin to form. The early Earth was a high-energy environment, under a Sun rich in ultraviolet radiation and with a turbulent atmosphere free from oxygen, and flashing with electrical storms. The chemical building-blocks that formed from these electrical discharges may have been concentrated in boiling mud-pools on the young Earth's numerous volcanoes.

## The complexity of life

The whole complexity of life can be thought of as a tree. The common root represents the origin of life with all the different forms originating from it. The largest branches that divide off from the trunk constitute the most fundamental varieties of living forms. These are the phyla (the plural of phylum). The thinnest twigs on the tree represent the separate species. There are other subdivisions between phylum and species. The whole system of classification, arranged from the general to the particular, runs as follows – Phylum / Class / Order / Family / Genus / Species. To find the common parents of two phyla we would have to go back hundreds of millions of years. On the other hand, two organisms that are different species of the same genus may have had common ancestors less than 100,000 years ago.



## What is Life?

Dogs, trees, and algae are alive; cars, stones, and computers are not alive. While some nonliving things may, for example, be able to move of their own accord, they all lack the fundamental property of life, the ability to create new copies of the living form: reproduction. Definitions of life do, however, encounter some problems with regard to viruses, which can grow and reproduce only when they have found their way into the cells of another living organism.

# Early life

The process of evolution over the first 3 billion years of life on Earth produced many remarkable new organisms. Only when these primitive algae formed large communities did they sometimes leave fossil records, pillars of limestone known as stromatolites. By around 2 billion years ago more complex "eukaryotic" plants had appeared. They had learned the advantage of sexual reproduction for creating greater diversity amongst offspring, hence speeding up evolution. Around 1 billion years ago the first multi-celled animals appeared, marking a breakthrough in the size and variety of life.

The first, tiny life-forms were almost defenseless. Around 590 million years ago at the beginning of the Cambrian period, a number of different types of organisms began to build themselves hard skeletons out of calcium carbonate. Sea urchins, for example, grew spherical armor plating, while mollusks developed shells similar to those found on seashores today. This innovation had the most dramatic impact on our knowledge of the history of life, because the hard parts have survived in the form of fossils.

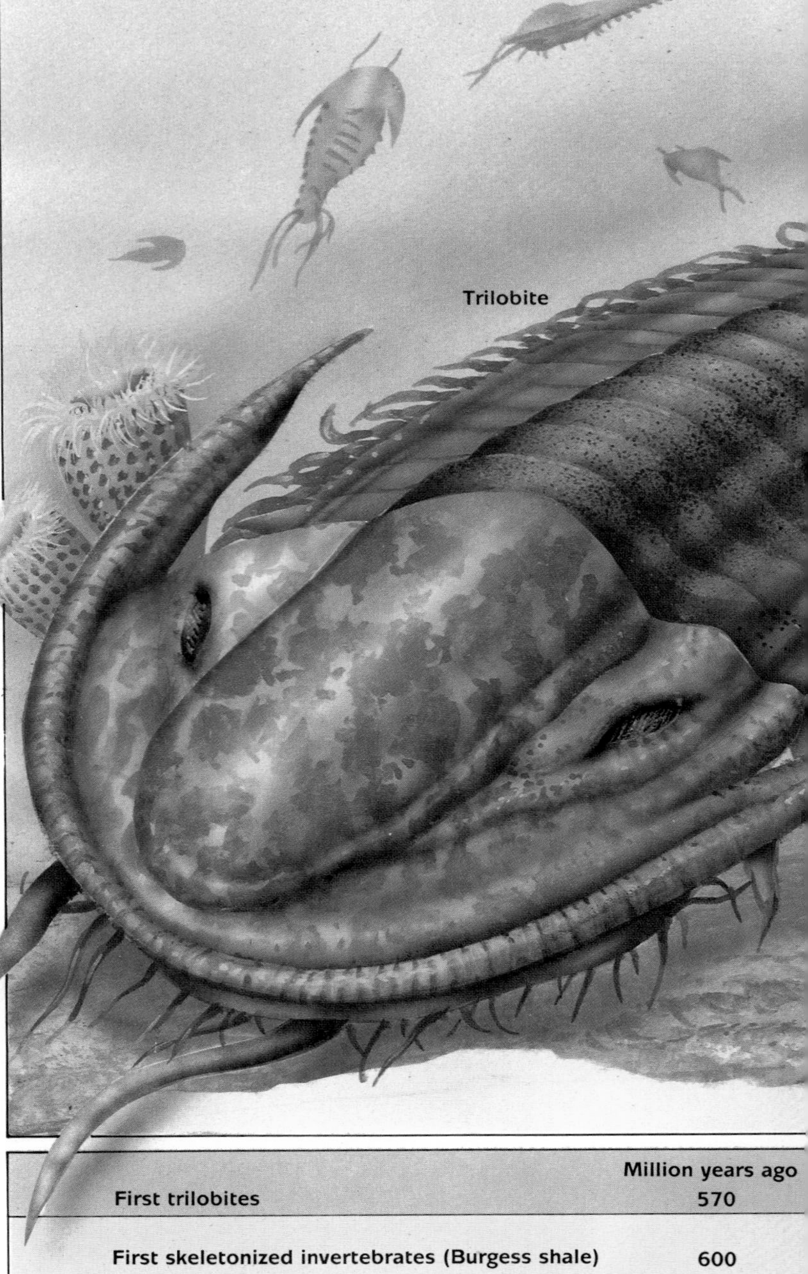


| | Million years ago |
|---|---|
| First trilobites | 570 |
| First skeletonized invertebrates (Burgess shale) | 600 |
| Ediacaran fauna (metazoans) | 650 |
| First eukaryotic plants | 1,400 |
| Cyanobacteria (blue-green algae) | 2,000 |
| First fossil remains of living organisms: stromatolites | 3,500 |
| Origin of the Earth | 4,600 |

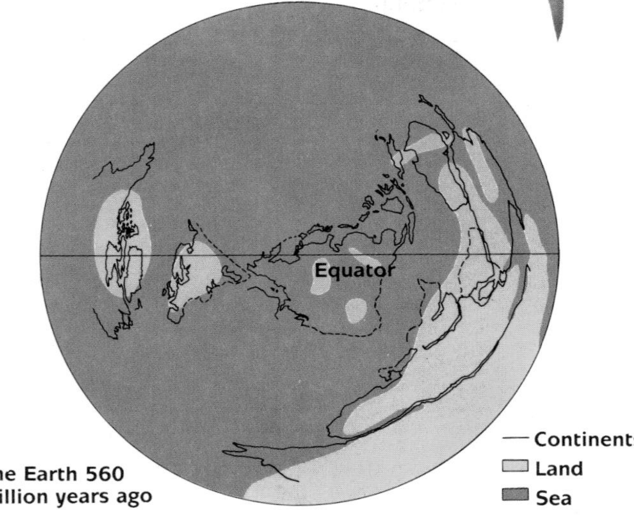


The Earth 560 million years ago



| Era | Precambrian 570 | (Million years ago) 500 | 430 | 395 | Paleozoic 345 | 280 | 225 | |
|---|---|---|---|---|---|---|---|---|
| Period | | Cambrian | Ordovician | Silurian | Devonian | Carboniferous | Permian | Tri |
| Geological time-scale | | | | | | | | |

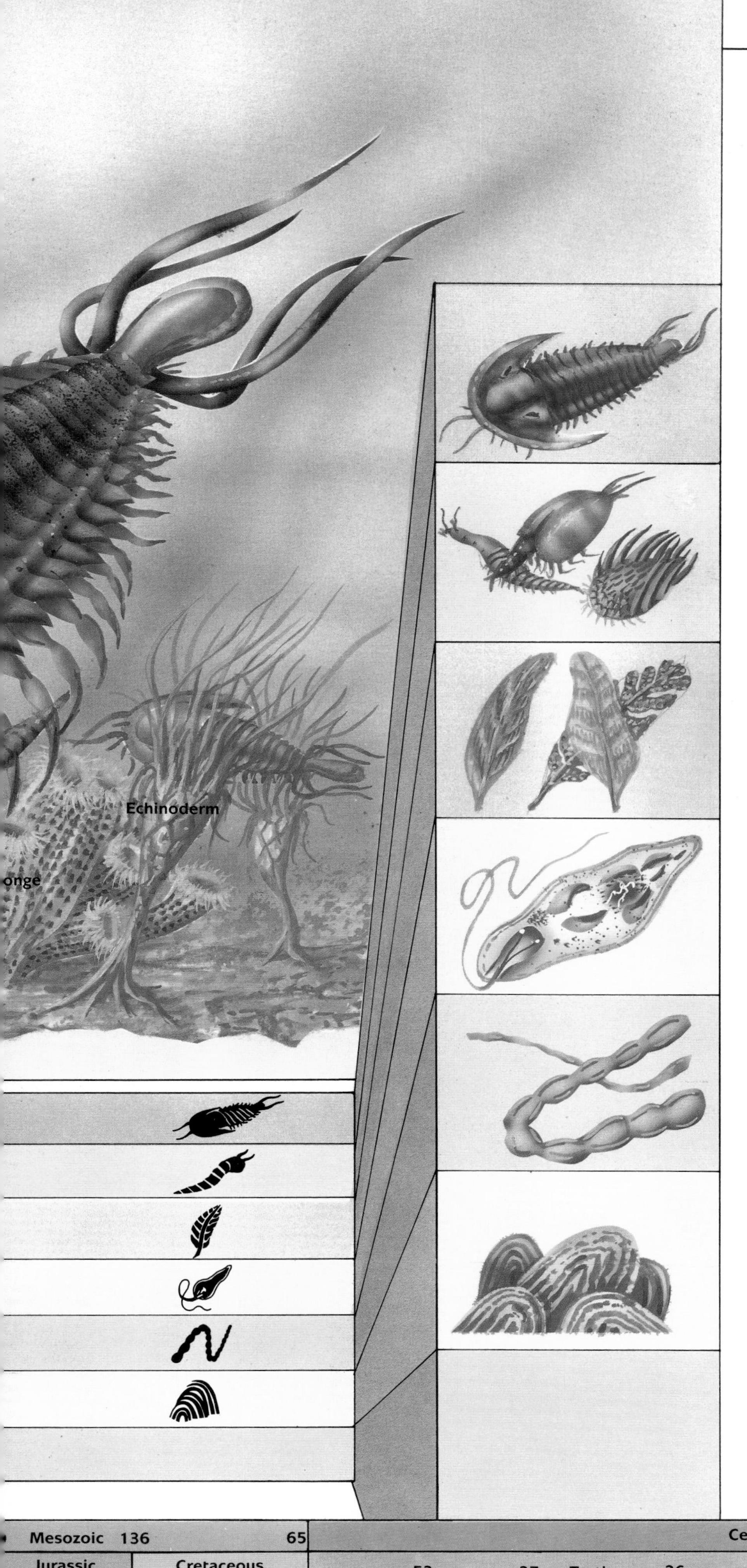


## Ediacara fossils

To discover what life was like immediately before this outburst of new forms we can look at the faint fossils in rocks from a period known as Ediacara. Nowhere are these fossils very abundant, but one can sometimes find the outlines of jellyfish, stranded on the beach, before becoming imprinted in the rock by a following tide, 600 million years ago. There are also feather-like forms from the sea floor, animal forms called polyps, which must have resembled modern sea pens.

## The Burgess shale

At Burgess Pass in British Columbia, Canada, can be found an outcrop of shale, formed from compressed muds and laid down on the ocean edge almost 550 million years ago. Within the fragile shale partings is preserved the best picture we have of the variety of hard- and even soft-bodied life-forms then in existence. Thirty million years after the Ediacara life-forms, the Burgess rocks were evidently teeming with extraordinary new organisms.

Many of the forms that we find as fossils in some of these Cambrian sediments have recognizable descendants (a few billion generations later) on modern beaches, sea floors, and swimming within the oceans. These include sea urchins, sponges, sea anemones, and worms. Other extraordinary organisms were experiments in design that failed. One of these had 15 segments, five eyes and a trunk ahead of its mouth; a second (christened Hallucigenia) had 14 legs, and seven tentacles, each possessing pincers.

## Trilobites

The group of animals that dominated marine life over the next 100 million years was the trilobites, creatures with broad heads and bodies covered with thin, rigid armor plating. The armor provided useful protection, but as the animal grew it had to discard its old coat and grow a new, bigger one. Hence many trilobite fossils are not dead animals but discarded armor.

Trilobites were the first organisms on Earth to evolve eyes. The eye was composed of up to 15,000 tiny calcite crystals offering some trilobites a well-rounded view of the world. Trilobites adapted to almost all marine environments, but those that moved to the deep dark seas lost their ability to see. Most trilobites scurried across the sea floor, like large marine woodlice.



Mesozoic 136 65 Cenozoic
Jurassic Cretaceous 53 37 Tertiary 26 5 1.6 Quaternary 0.01
Paleocene Eocene Oligocene Miocene Pliocene Pleistocene Holocene

# Life conquers the land

For the first 4 billion years of the Earth's history the land was as barren as the Moon. With nothing to bind the particles of weathered rock, mountains eroded rapidly, rivers were muddy with sediment and enormous dust storms swept the continents.

Around 420 million years ago plant cells developed the ability to protect themselves within a waxy skin. The plant could now carry its own reduced water supplies and was thus able to colonize land far from the water's edge. The earliest plants simply lay on the surface of wet ground. Away from marshes plants developed roots that could grow downward, to tap water trapped between soil particles below the surface. The development of roots required the creation of pipework within the plant's structure, rigid enough to prevent blockages. This in turn allowed plants to grow tall.

A race for the sky had begun. As the tallest plant could shade its neighbors and stunt their growth, taller and taller trees evolved.

1 **Club moss**
2 **Horsetail**
3 **Fern**
4 ***Eryops***
5 **Dragonfly**

### Plant reproduction

The earliest plants of the forest – the club mosses, horsetails and ferns – reproduced through an intermediate stage in which a tiny, moisture-rich plant, the thallus, grew on the forest floor into which it dropped its spores. Ferns, club mosses, and horsetails still require moist conditions for their survival. However, some 350 million years ago new reproductive strategies evolved that raised reproduction above the ground and the need for water.

The most successful of these innovations occurred among trees, in which the wind carried the male spores (pollen), a few of which chanced to land on clusters of female eggs wrapped in woody material. The pollen developed a tube that over a year grew down to unite the spore with the egg. The fertilized egg grew into a food-rich waterproof package (the seed), that could survive several years before conditions were appropriate for growth. These arrangements of eggs are known as

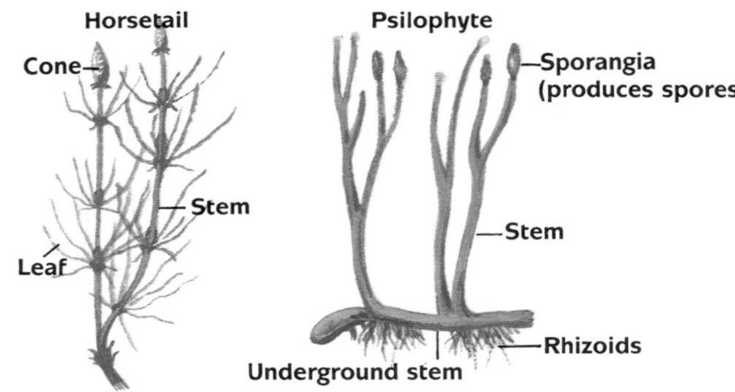


cones, and the early trees on which they developed, the conifers, became particularly successful.

**The Earth 320 million years ago**



| Era | Precambrian 570 | (Million years ago) 500 | 430 | 395 | Paleozoic 345 | 280 | 225 |
|---|---|---|---|---|---|---|---|
| Period | | Cambrian | Ordovician | Silurian | Devonian | Carboniferous | Permian | Tria |
| Geological time-scale | | | | | | | | |

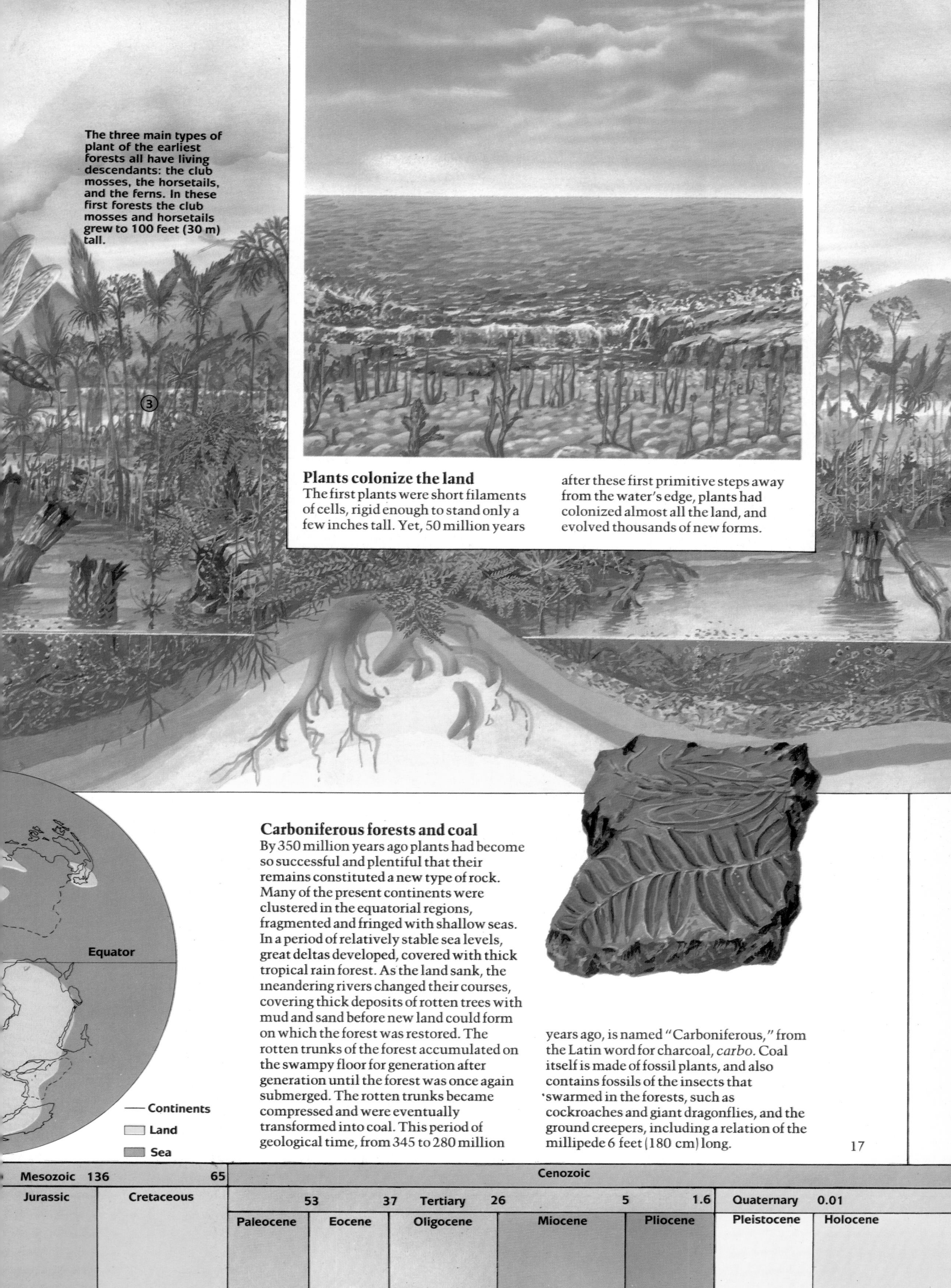


The three main types of plant of the earliest forests all have living descendants: the club mosses, the horsetails, and the ferns. In these first forests the club mosses and horsetails grew to 100 feet (30 m) tall.

## Plants colonize the land

The first plants were short filaments of cells, rigid enough to stand only a few inches tall. Yet, 50 million years after these first primitive steps away from the water's edge, plants had colonized almost all the land, and evolved thousands of new forms.

## Carboniferous forests and coal

By 350 million years ago plants had become so successful and plentiful that their remains constituted a new type of rock. Many of the present continents were clustered in the equatorial regions, fragmented and fringed with shallow seas. In a period of relatively stable sea levels, great deltas developed, covered with thick tropical rain forest. As the land sank, the meandering rivers changed their courses, covering thick deposits of rotten trees with mud and sand before new land could form on which the forest was restored. The rotten trunks of the forest accumulated on the swampy floor for generation after generation until the forest was once again submerged. The rotten trunks became compressed and were eventually transformed into coal. This period of geological time, from 345 to 280 million years ago, is named "Carboniferous," from the Latin word for charcoal, *carbo*. Coal itself is made of fossil plants, and also contains fossils of the insects that swarmed in the forests, such as cockroaches and giant dragonflies, and the ground creepers, including a relation of the millipede 6 feet (180 cm) long.

# The first vertebrates

The last major animal order to appear was the vertebrates, animals with backbones. An internal skeleton allowed numerous adaptations into strong, large and fast-moving animals. Small bone-armored primitive fish ancestors began to develop 550 million years ago. In time these evolved gills, thin blood vessels in slits adjacent to the throat, through which dissolved oxygen could be extracted from water. By around 450 million years ago true fish had appeared, complete with jaws and lateral fins. There were two distinct types. The first group, which replaced all the bone with lighter elastic cartilage, includes among its modern descendants the rays and sharks. The other group of fish kept their bones and developed air sacs for better buoyancy in the sea.

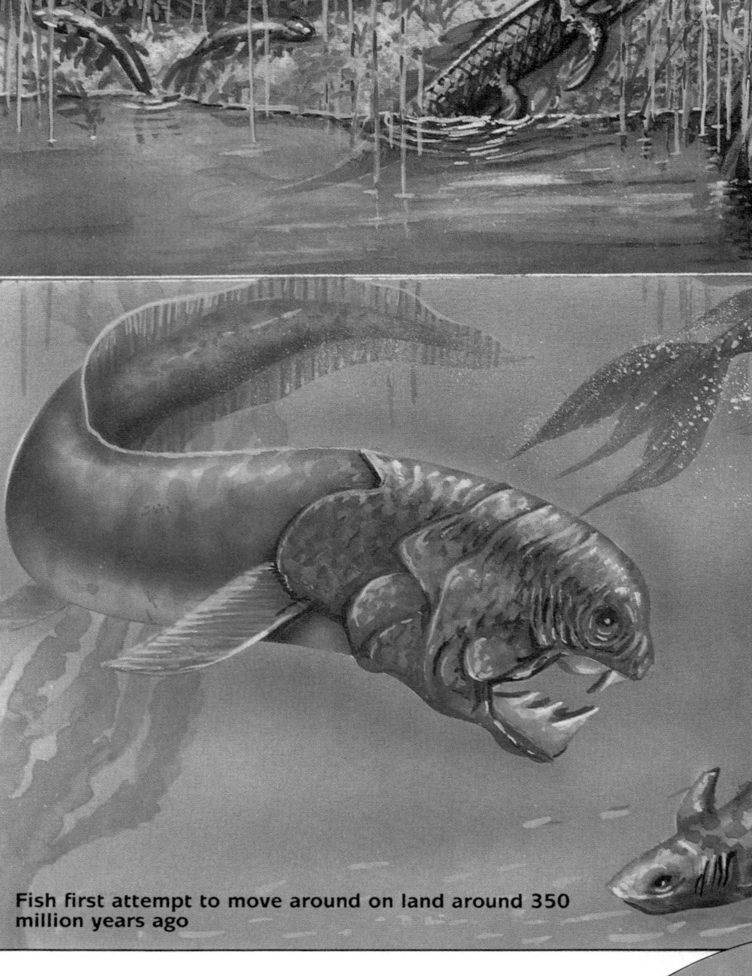

**Fish first attempt to move around on land around 350 million years ago**

*Dimetrodon*

**From around 270 million years ago, during the Permian period, the king of the jungle was a reptile, *Dimetrodon*, which grew to up to 6 feet (2 m) in length and had a great curve of skin all along its back, held up by thick spines. This strange spiny sail may have helped the cold-blooded animal to control its temperature, allowing blood to be rapidly warmed by the sun, or cooled when fanned in the breeze.**

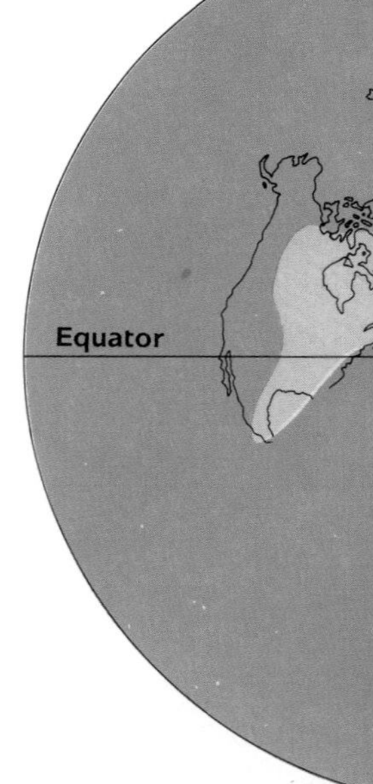


### Early reptiles

Eggs with shells developed around 300 million years ago, and marked the emergence of the reptiles. By the end of the Permian period strange, thick-skulled herbivorous reptiles had become the top land animals. The first crocodiles appeared around 230 million years ago, some with heads 3 feet (1 m) long, filled with saw-teeth. By 200 million years ago, these advanced reptiles had displaced the thick-skulled reptiles from many of their habitats, and had begun to eat plants. The dominance of these new animals, the dinosaurs, was to endure for 130 million years.



| Era | Precambrian 570 | (Million years ago) 500 | 430 | 395 | Paleozoic 345 | 280 | 225 | |
|---|---|---|---|---|---|---|---|---|
| Period<br>Geological time-scale | | Cambrian | Ordovician | Silurian | Devonian | Carboniferous | Permian | Tri |

## Vertebrates leave the water

Fish first attempted life on land around 350 million years ago. These first land-living creatures with limb-like fins are known as amphibians and still relied on water for reproduction. Modern amphibian descendants include frogs, toads, and newts. For the first 50 million years of life on land all vertebrates were amphibian, returning to the water to breed. Today, amphibian eggs hatch as fish-like tadpoles; only adults are fully adapted to life out of water. It was only around 300 million years ago that plant-eating vertebrates appeared, allowing an explosion in the numbers and variety of land animals.

## Lizards take to the sea

Some reptiles returned to terrorize the oceans. In the plesiosaurs the four limbs became powerful flippers, and the elongated neck supported a narrow head with a sharp set of teeth to snatch fish. Like the modern turtle, the plesiosaur probably returned to a beach to lay its eggs. In contrast, the icthyosaur had no neck, a broad streamlined body and a long tail. The limbs had become flippers and the eyes had adapted to chasing prey at speed through the water. Both plesiosaurs and ichthyosaurs were carnivores, feeding on the plentiful fish, and both flourished for almost 200 million years.

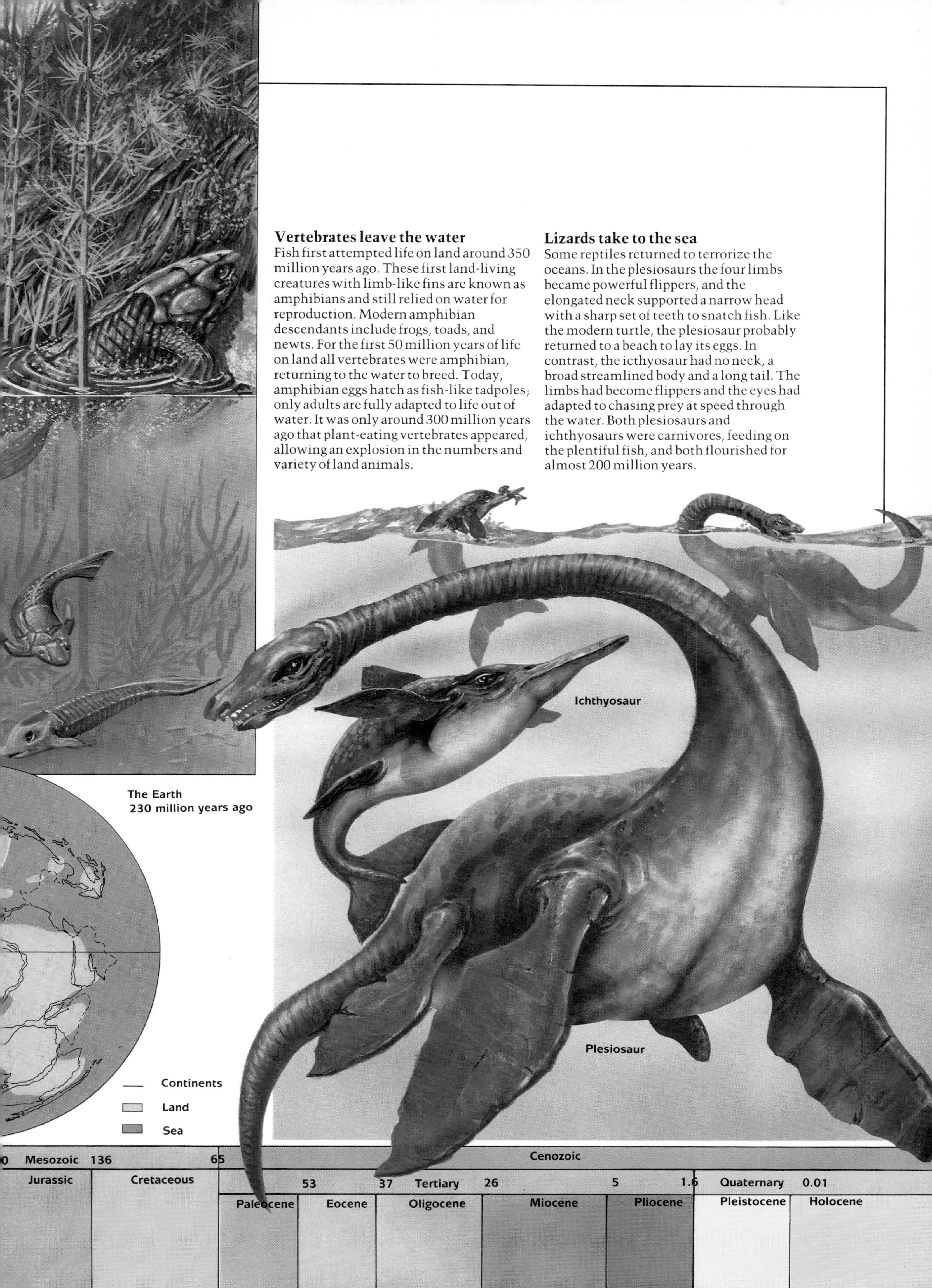

THE HISTORY OF LIFE

# The dispersal of Pangaea

Between 400 and 250 million years ago, in a series of gigantic collisions, the separate continents welded into a single supercontinent – Pangaea. Around 200 million years ago, it began to break up; huge thermal upwellings in the underlying mantle dragged at the crust, splitting it along rifts that eventually developed into ocean basins. While the northern continents – principally Eurasia and North America (known as Laurasia) – remained in close contact and continued to share populations, the southern half of Pangaea, or – South America, Africa, India, Australia and Antarctica (known collectively as Gondwana) – voyaged independently.

The history of the fragmentation of the Pangaea supercontinent has had an immense impact on the evolution of life. As the continents drifted apart, so the animals and plants evolved into distinct new species, while many original species became extinct. Most new species had evolved in the tropics, and as the Gondwana continents moved slowly around the temperate latitudes of the southern hemisphere, they preserved older life-forms that had formerly lived in Laurasia.

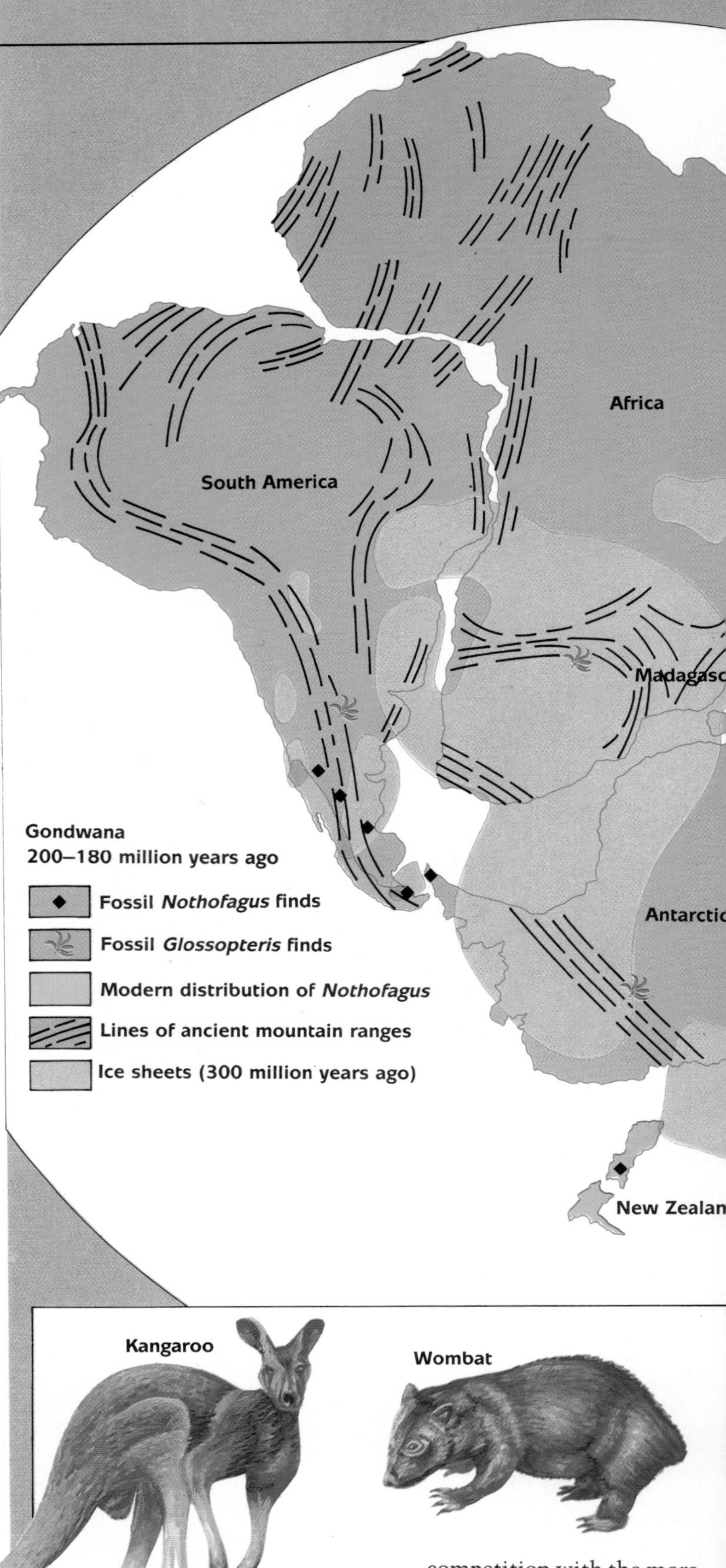

**The tuatara**
In New Zealand, of all the Gondwanan reptiles, only the tuatara survived. Some 100 million years ago, New Zealand was closer to the South Pole and temperatures were too cold for all reptiles except for the tuatara.

**The marsupials**
Everywhere apart from South America and Australia marsupials have become extinct, because they failed in competition with the more advanced "placental" mammals. However, these had not emerged before Australia had become isolated: hence all native Australian mammals are marsupial.

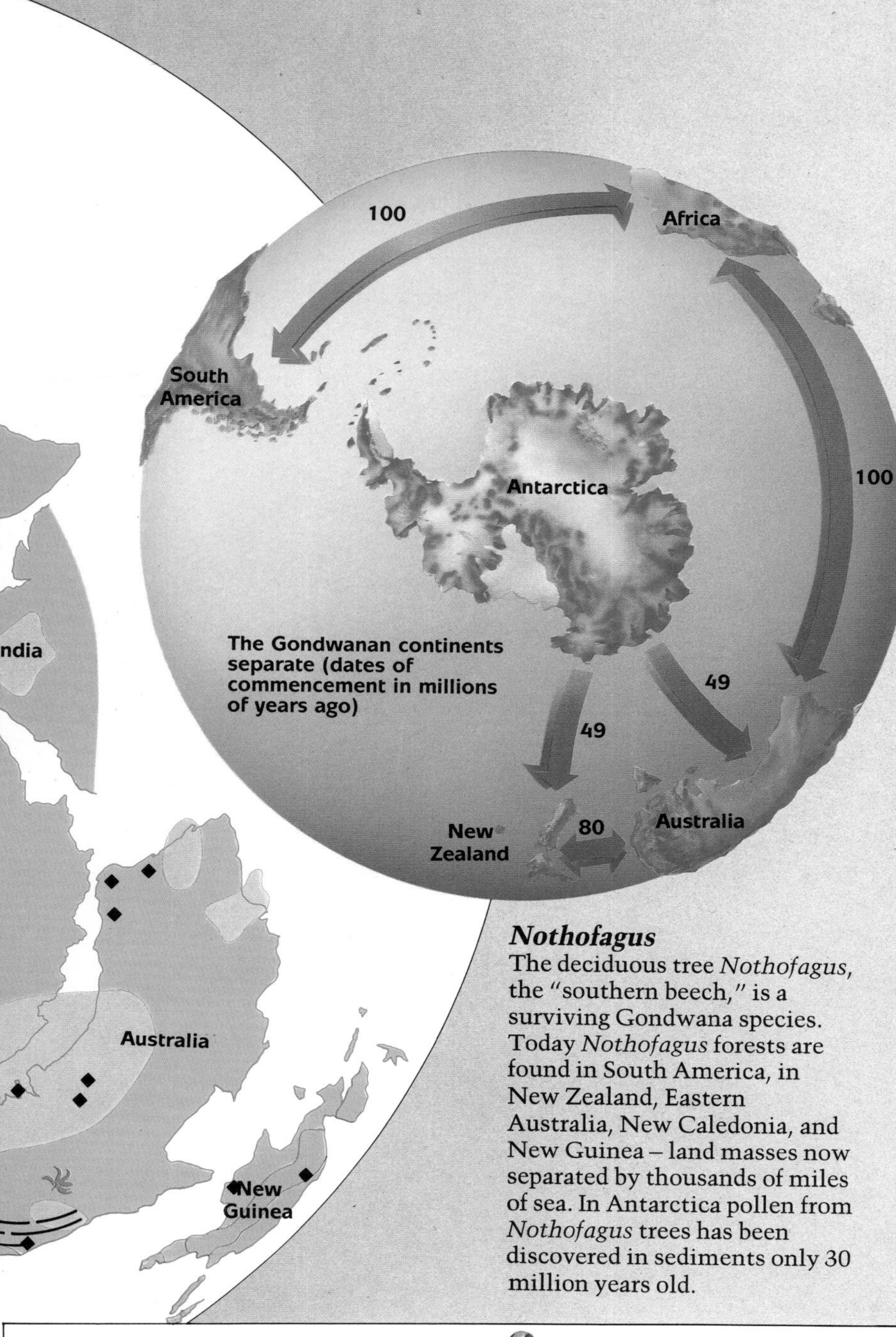


## Pangaea falls apart

Pangaea first began to fall apart between Laurasia and Gondwana at what is now the mid-Atlantic, separating the eastern coast of the United States from the northwestern edge of Africa. This separation began around 180 million years ago. The Atlantic Ocean began to form, first in the far south, around 135 million years ago. By 100 million years ago it extended all the way to the Bay of Biscay as well as between Greenland (then connected to the northwestern edge of Europe) and North America. Around 55 million years ago Greenland itself began separating from Europe.

The continents of Gondwana also began to separate to the east of Africa. First India, Africa, and Australia (attached to Antarctica), all parted company around 180 million years ago. Madagascar separated from Africa about 100 million years ago, while New Zealand abandoned Australia 80 million years ago. It was not until around 50 million years ago that Australia separated from Antarctica.

The story of the floating continental "rafts" of Gondwana, and exactly when they separated from one another is not always easy to unravel, but the best source of evidence comes from "living fossils," direct descendants of Gondwanan plants and animals that have undergone little change, such as the marsupials and ratites.

### *Nothofagus*

The deciduous tree *Nothofagus*, the "southern beech," is a surviving Gondwana species. Today *Nothofagus* forests are found in South America, in New Zealand, Eastern Australia, New Caledonia, and New Guinea – land masses now separated by thousands of miles of sea. In Antarctica pollen from *Nothofagus* trees has been discovered in sediments only 30 million years old.

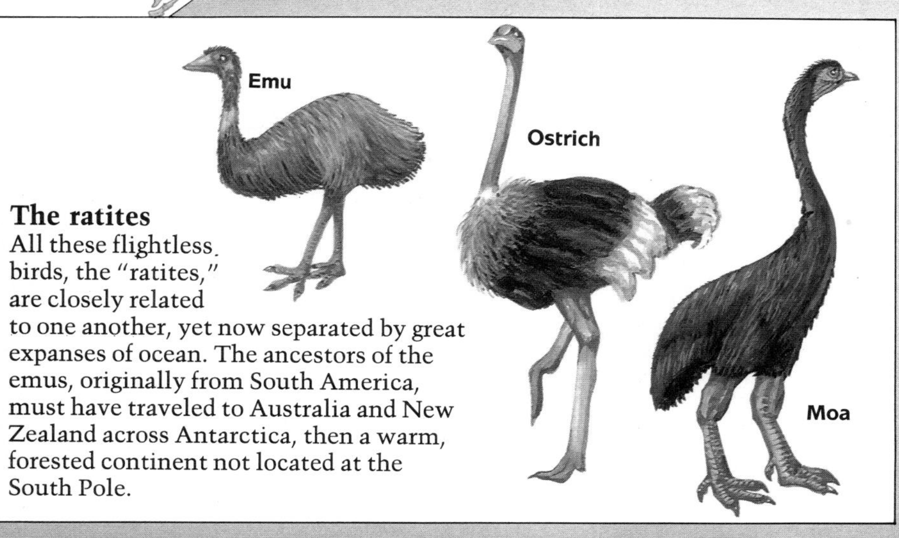


### The ratites

All these flightless birds, the "ratites," are closely related to one another, yet now separated by great expanses of ocean. The ancestors of the emus, originally from South America, must have traveled to Australia and New Zealand across Antarctica, then a warm, forested continent not located at the South Pole.

### *Glossopteris*

*Glossopteris*, a fern-like plant, was extremely abundant 250 million years ago. Its fossilized leaves are common in rocks of India, Australia, Borneo, South Africa, South America, and Antarctica.

THE HISTORY OF LIFE

# The age of the dinosaurs

The fight to rule all the Earth's jungles, deserts, and swamps was won around 200 million years ago, by a very remarkable group of animals called the dinosaurs. All those animals that had formerly ruled the land, such as the early amphibians and lizards, as well as the more recently evolved reptiles such as the crocodiles, survived the rule of the dinosaurs simply by staying out of their way. There were two strategies: some animals (like the mammals and lizards) had to stay small in size so that they could hide in burrows and crevices, or else (like the crocodiles) they lived chiefly in the rivers, lakes, and in the sea.

Stegosaurus

### Dinosaur species

Over a period of 130 million years, different species of dinosaur emerged to command the various environments. This evolution was an arms race – a carnivore had to become ever better armed to ensure its survival. The early *Stegosaurus* protected itself with a set of eight enormous spikes at the end of its massive rigid tail; it also had a series of plates along its spine, that could be lowered as defensive body-armor. The *Stegosaurus* had to defend itself against the two most common giant meat-eaters, *Allosaurus* and *Ceratosaurus*, which grew up to 30 ft (9 m) long and weighed 1 to 12 tons.

Halfway through the age of the dinosaurs, at the end of the Jurassic geological period, the stegosaurs almost disappeared. In their place emerged new armored dinosaurs, the four-ton nodosaurs, covered with bony armor-plating and enormous spines. At this time in the deltas of the western United States lived the *Ankylosaurus*. It had a thick club at the end of its long flexible tail which it could whip around to smash into an attacker. The giant *Brontosaurus* depended on its huge size to discourage aggressors as it used its 70-foot (21 m) length to browse among the treetops.

The most powerful defensive armor of all was to be found on the *Triceratops*, which lived towards the end of the reign of the dinosaurs. It defended itself with its colossal armored head, a fearsome weapon used at speed. *Triceratops* had the largest skull ever to appear on land, up to 8 ft (240 cm) in length and more than 3 ft (90 cm) broad. On top of this was a set of forward-projecting horns over 3 ft (90 cm) long; a collar of rugged bone fringed the neck. *Triceratops* had need of this protection, because at this period lived the most dangerous carnivore ever: *Tyrannosaurus Rex*, 20 ft (6 m) high and five tons of fighting machine, with the most powerful jaws in all prehistory. Almost like a kangaroo its forearms had shrunk to insignificance.

Nodosaurus

Ankylosaurus

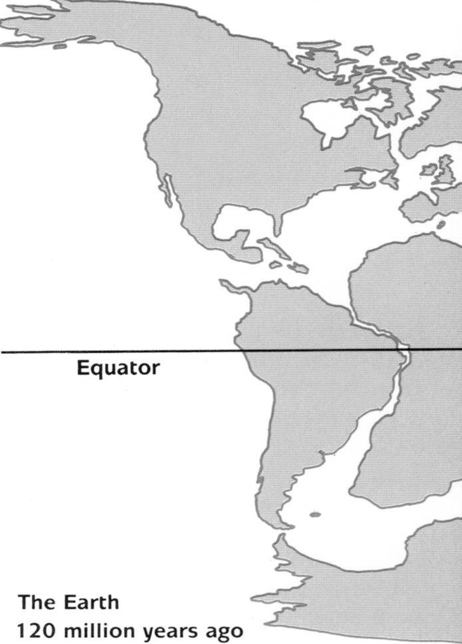


The Earth
120 million years ago



| Era | Precambrian 570 | (Million years ago) 500 | 430 | 395 Paleozoic | 345 | 280 | 225 | |
|---|---|---|---|---|---|---|---|---|
| Period<br>Geological time-scale | | Cambrian | Ordovician | Silurian | Devonian | Carboniferous | Permian | Tria |

Ceratosaurus

Allosaurus

Archaeopteryx

## Dinosaurs take to the air

Dinosaurs were spectacularly successful on land, but failed to adapt to life in rivers or the sea. Instead they adapted into a world previously colonized only by small insects – the air. The first flier was the raven-sized *Pterodactyl*. This had a long jaw filled with sharp teeth, thin-walled hollow bones like those of modern birds, and a scaly membrane wing stretched over its elongated fingers. In the Cretaceous period, the long-tailed *Pterodactyl* was replaced by huge short-tailed forms.

In 1861 in the Jurassic limestone quarry at Solenhofen, West Germany, an almost complete skeleton was found of a small pigeon-sized dinosaur with thick feathered wings and a long bony tail. This specimen, *Archaeopteryx*, proved to be the missing link between modern birds and the dinosaurs. Modern toothless birds only appeared at the end of the Cretaceous period.

Triceratops

Tyrannosaurus Rex

Pterodactyl

ntosaurus

## Dinosaur remains

Dinosaur history is best preserved in the midwestern US, China, and Mongolia. Nowhere has the whole record survived, as ideal locations for preserving dinosaur skeletons have shifted continually. In the Jurassic, when primitive dinosaurs like *Brontosaurus* thrived, the western US comprised meadows and shallow lakes, while in the Cretaceous the same region was mostly swampy deltas.

| 0 Mesozoic 136 | | 65 Cenozoic | | | | | | | |
|---|---|---|---|---|---|---|---|---|---|
| Jurassic | Cretaceous | 53 | 37 Tertiary | 26 | 5 | 1.6 | Quaternary 0.01 | |
| | | Paleocene | Eocene | Oligocene | Miocene | Pliocene | Pleistocene | Holocene |

# The rise of the mammals

The earliest true mammals appeared around 180 million years ago. Descendants of the earliest reptile-mammals survive today in parts of Australia, and include the duck-billed platypus, which lays eggs like a reptile but suckles its young like a mammal. Mammals are warm-blooded, hairy creatures who sustain their body heat independent of the external temperature. To achieve this requires a high level of energy and they had to be efficient at finding and consuming food. Even after 100 million years early mammals grew no larger than a rat, and may have been nocturnal to avoid competition with the cold-blooded reptiles, who could hunt only by day. They survived on their wits: weight for weight they had larger brains than the dinosaurs. Sixty-five million years ago, with the death of the dinosaurs, the mammals were suddenly presented with a great range of opportunities. In only 15 million years they had evolved into the whole range of modern mammals.

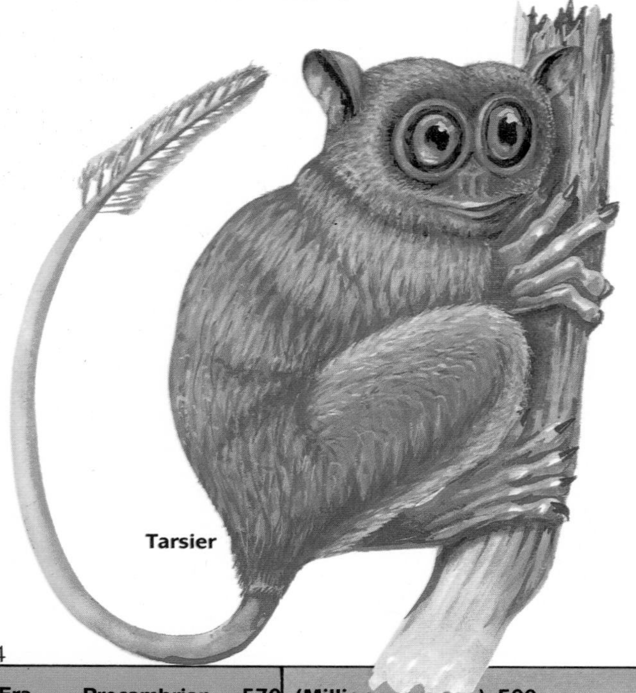

Tarsier

**The rise of the primates**
Primitive monkeys, the lemurs, evolved around 50 million years ago. When Madagascar separated from East Africa some 30 million years ago, it carried a population of lemurs that has survived to the present day. Elsewhere only a few species, such as the tarsier and bushbaby, survived. In Africa some species of monkey moved onto the ground. These creatures already possessed an enlarged brain for stereoscopic vision. Among the ground-living primates was a wily ape, which over the period of a few million years evolved into *Homo sapiens*.

| Era | Precambrian | 570 (Million years ago) 500 | | 430 | 395 Paleozoic | 345 | 280 | 225 |
|---|---|---|---|---|---|---|---|---|
| Period | | Cambrian | Ordovician | Silurian | Devonian | Carboniferous | Permian | Tri |
| Geological time-scale | | | | | | | | |

## Mammals take to the air

Mammals first took to the air in the forest where, like the dinosaurs before them, the adaptation assisted in moving from tree to tree. By 50 million years ago true bats had evolved with a wing of skin stretching back to the body from the second finger, leaving only the thumb as an all-purpose claw.

Bats, like their mammalian ancestors at the time of the dinosaurs, are the "flying night shift," an acknowledgment that there was little to gain in competition with the descendants of the dinosaurs: the birds. Emitting bursts of very high-pitched squeals, bats orient themselves from the echo and navigate by sonar. Some species are night hunters of fruit, flowers or even birds and fish, in place of their original insect diet.

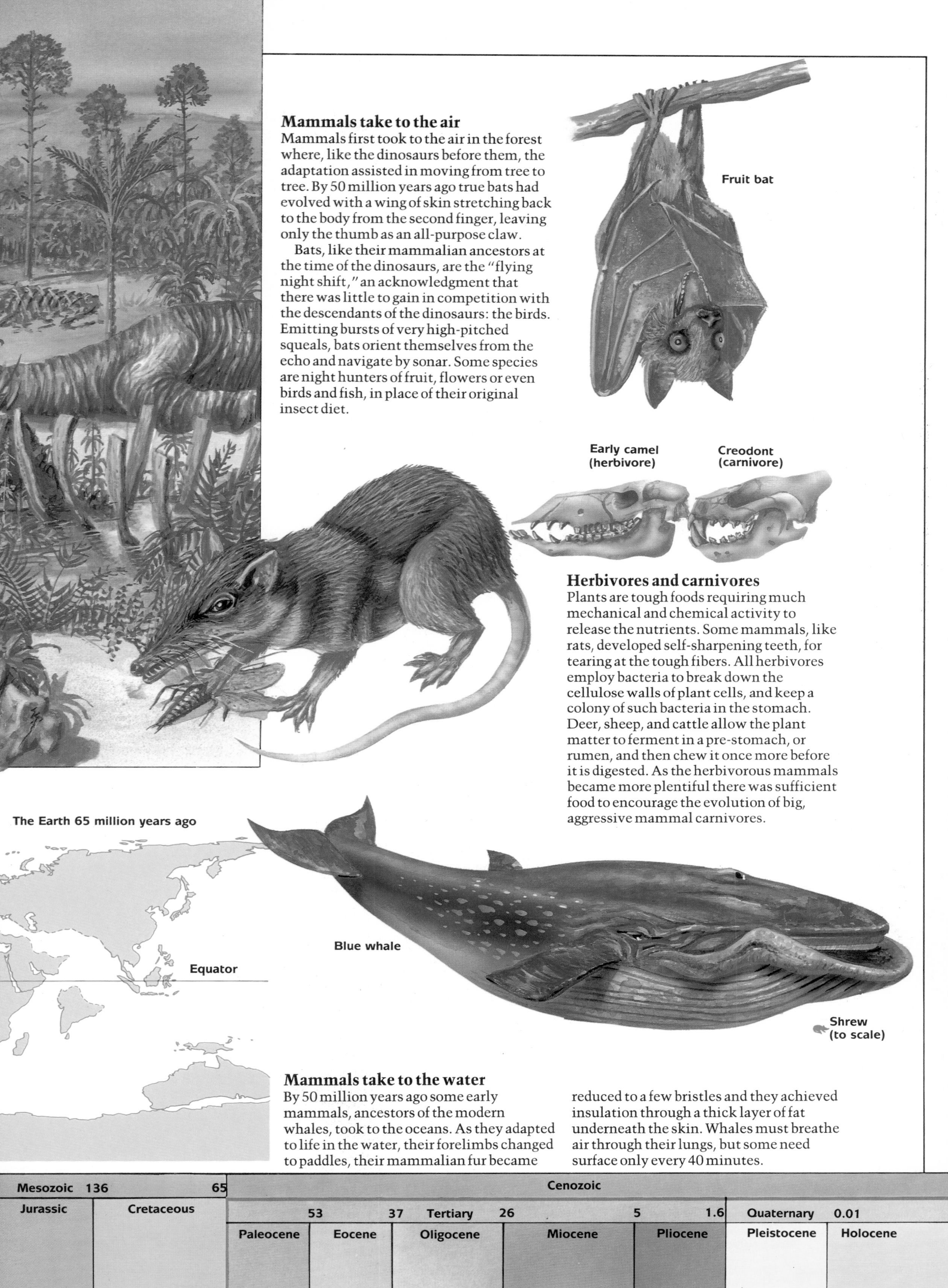


## Herbivores and carnivores

Plants are tough foods requiring much mechanical and chemical activity to release the nutrients. Some mammals, like rats, developed self-sharpening teeth, for tearing at the tough fibers. All herbivores employ bacteria to break down the cellulose walls of plant cells, and keep a colony of such bacteria in the stomach. Deer, sheep, and cattle allow the plant matter to ferment in a pre-stomach, or rumen, and then chew it once more before it is digested. As the herbivorous mammals became more plentiful there was sufficient food to encourage the evolution of big, aggressive mammal carnivores.

## Mammals take to the water

By 50 million years ago some early mammals, ancestors of the modern whales, took to the oceans. As they adapted to life in the water, their forelimbs changed to paddles, their mammalian fur became reduced to a few bristles and they achieved insulation through a thick layer of fat underneath the skin. Whales must breathe air through their lungs, but some need surface only every 40 minutes.

| Mesozoic 136 | 65 | Cenozoic | | | | | | | |
|---|---|---|---|---|---|---|---|---|---|
| Jurassic | Cretaceous | 53 | 37 | Tertiary 26 | 5 | 1.6 | Quaternary 0.01 | | |
| | | Paleocene | Eocene | Oligocene | Miocene | Pliocene | Pleistocene | Holocene | |

# The Ice Ages

About 10 million years ago a permanent ice cap developed over the Antarctic continent. By 2.5 million years ago icebergs were passing into the North Atlantic. This was the beginning of the first Ice Age for almost 300 million years. There have been around seven separate cold periods in the past 700,000 years, and on at least three occasions ice caps, up to 2 miles (3.2 km) thick, have developed over northern Europe and North America. The last Ice Age lasted from about 40,000 to 14,000 years ago.

During the intervening "interglacials" the climate was at times even warmer than it is today, and elephants and hippos flourished in western Europe. While the continental ice sheets destroyed much wildlife they did help to create new habitats around the margins of the continents. Millions of cubic miles of water from the oceans were turned into ice, lowering the seas by some 400 ft (120 m). All around the world the shallow seas of the continental shelves became wide tracts of lowland.

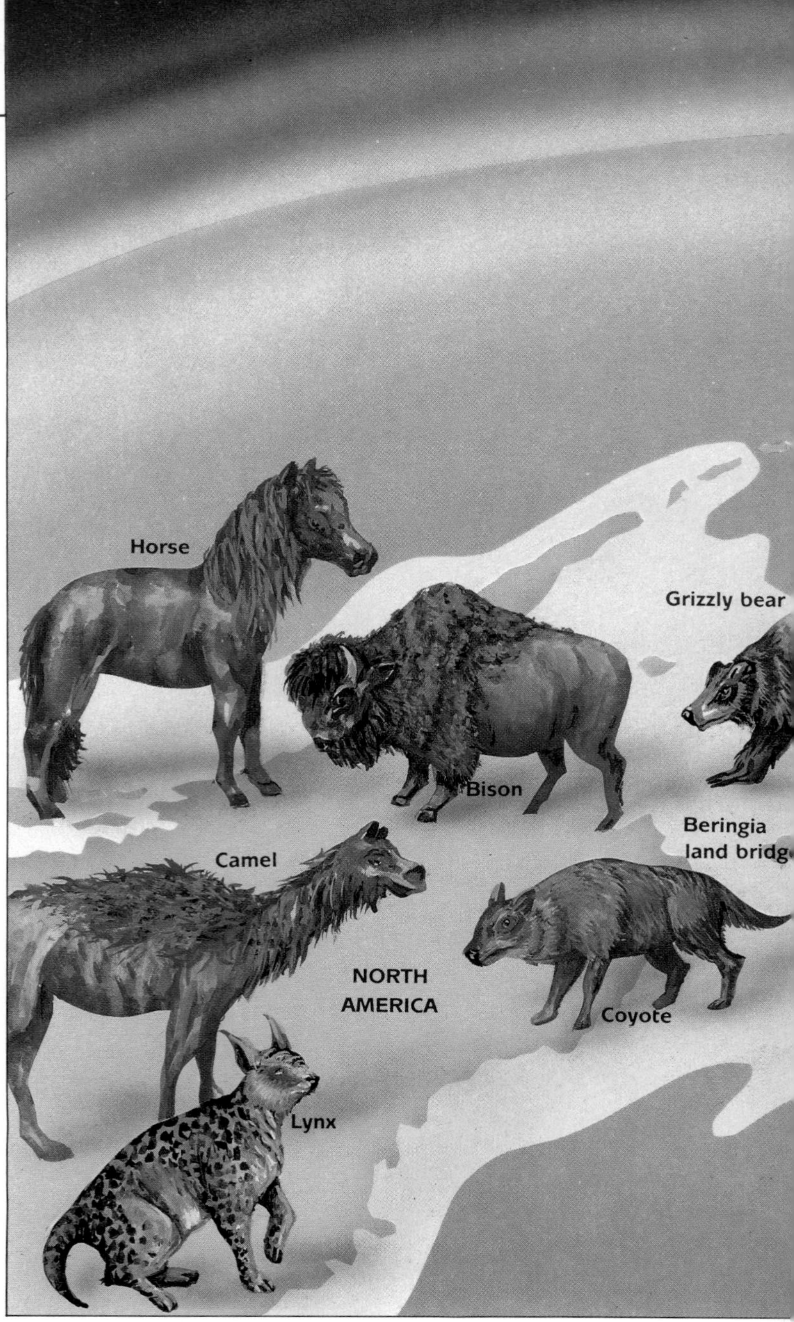


The sea contours give an impression of the shape of the "British Isles" before rising sea levels made them islands.

100 m
Strawberry tree
Ireland
150 m
Great Britain
100 m
Grass snake

### Recolonizing the barren lands

After the temperatures in the northern hemisphere began to rise again, about 13,000 years ago, plants and animals recolonized those barren areas abandoned by the ice. In some regions this was a race against rising sea levels. In Great Britain and Ireland the distribution of species compared to those on the continent records the pattern of this invasion. Britain became an island around 8,000 years ago, and Ireland probably sometime before this. As a result, some less hardy species never made it to Ireland: the list includes wild lime trees and snakes. Conversely, a few indigenous species of wild plants (including the strawberry tree) arrived in Ireland via the continental shelf from northern Spain, but never got as far as Britain. Throughout northern Europe and northern North America the range of species is very impoverished compared to that which existed before the Ice Age.

## Beringia

From about 50 to 10 million years ago there was a broad, forested land bridge connecting Asia with North America across "Beringia," the region around the present Bering Strait. As the climate cooled this deciduous forest, retreated south into what is now Japan and China to the west and the southeastern United States to the east, to be replaced by the great northern coniferous forest, known as the taiga. Eventually Asia and North America became separated by sea, and the animal populations no longer moved between the continents. At the height of the Ice Age the sea level fell and Beringia became dry land once again. Across this grassy tundra came herds of mammals, almost all traveling for the first time from Asia into North America, including musk ox, moose, brown bear, and wolf, species that are found today in Canada. Other species traveled farther south into what is now the United States. They included bison, grizzly bear, coyote, and bobcat. Some of these emigrants, such as the mammoth and the saber-tooth tiger did not survive after the retreat of the glaciers.

One of the last ice age migrants from Asia to America, some 17,000 to 19,000 years ago, was *Homo sapiens*. All the native peoples of North and South America are descendants of these intrepid refugees from northeast Asia.

## Diversity in the rain forest

The expanded Arctic regions that accompanied the Ice Ages had a significant impact on global climate, even in equatorial regions. In the area now covered by the Amazonian rain forest, wet periods have alternated with dry periods, when the forest shrunk into a few widely separated wooded "islands." On each of these islands, the species tended to evolve along separate paths. After a few tens of thousands of years they could no longer breed with their former cousins on other forest islands. Hence, when the climate became wet again, and the forest islands grew to form one enormous rain forest, the number of distinct species multiplied. The pattern of climate change in the past 2 million years is thought to have been a principal cause of the rich diversity of species (both plant and animal) in the rain forest.

EVOLVING ENVIRONMENTS

# Mid-ocean ridges

Running the length of the exact center of the Atlantic Ocean is a range of mountains lying for almost all its extent more than a mile underwater. This ridge marks a line of rifts, giant cracks, and volcanic peaks, beneath which hot mantle rock rises, and as it rises, begins to melt. The liquid rock, or magma, generally freezes below ground, but some pours out on the seafloor, where it forms large magma bubbles. These become chilled by the water, and collapse into pillow shapes.

The mid-ocean ridge marks the location where two plates are separating from one another; the rising mantle is simply arriving to fill the gap. If the water were drained away to expose the spreading ridge, it would appear to have a central rift, in the midst of the mountains. It is along this rift that the plates move apart and the volcanic activity is concentrated.

Spreading ridges widen at rates of a few inches a year, the rate at which fingernails grow.

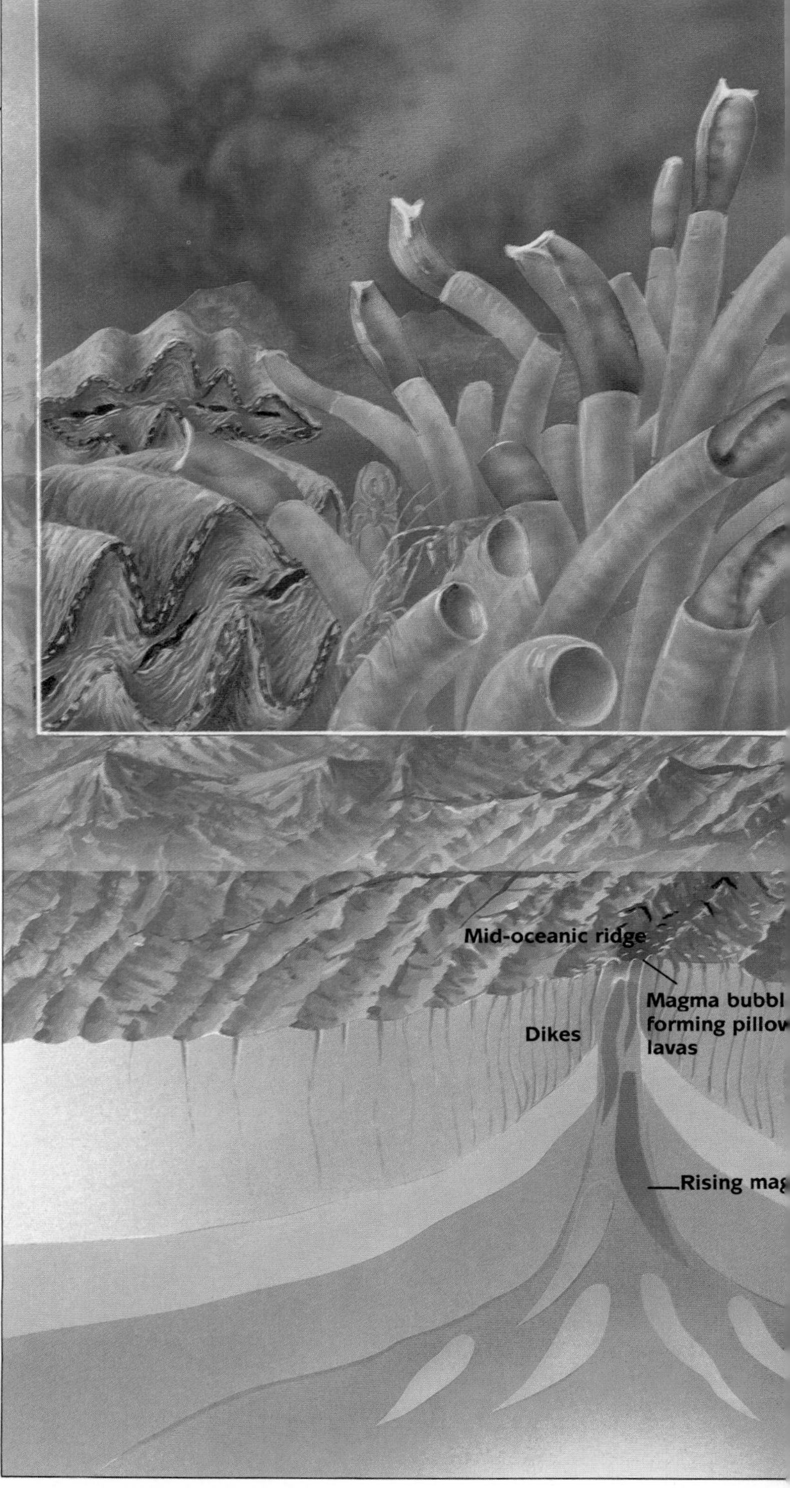


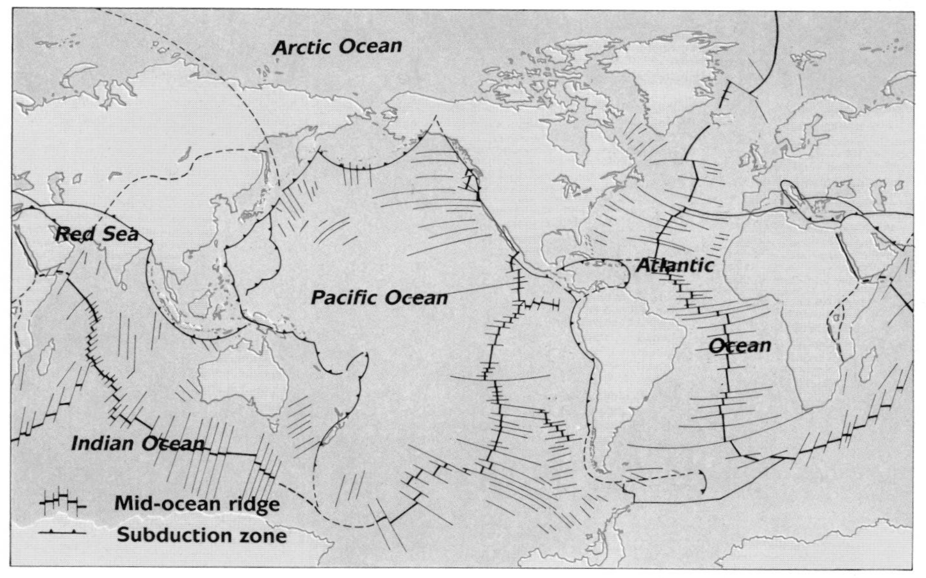


**Spreading ridges of the world**
The mid-ocean spreading ridge can be followed beyond the Atlantic Ocean. To the north it crosses the center of the Arctic Ocean, eventually ending against the coast of northeastern Siberia. In the south it passes eastward into the Indian Ocean, with one offshoot running up into the Red Sea.

Only at a few locations on Earth does the spreading ridge rise above sea level. The most important location is Iceland, where one can stand astride two plates. Iceland is built entirely from geologically recent eruptions predominantly of basaltic magma. The island continues to expand when, in sudden episodes of volcanic activity, fresh supplies of magma fill and widen vertical cracks known as dikes.

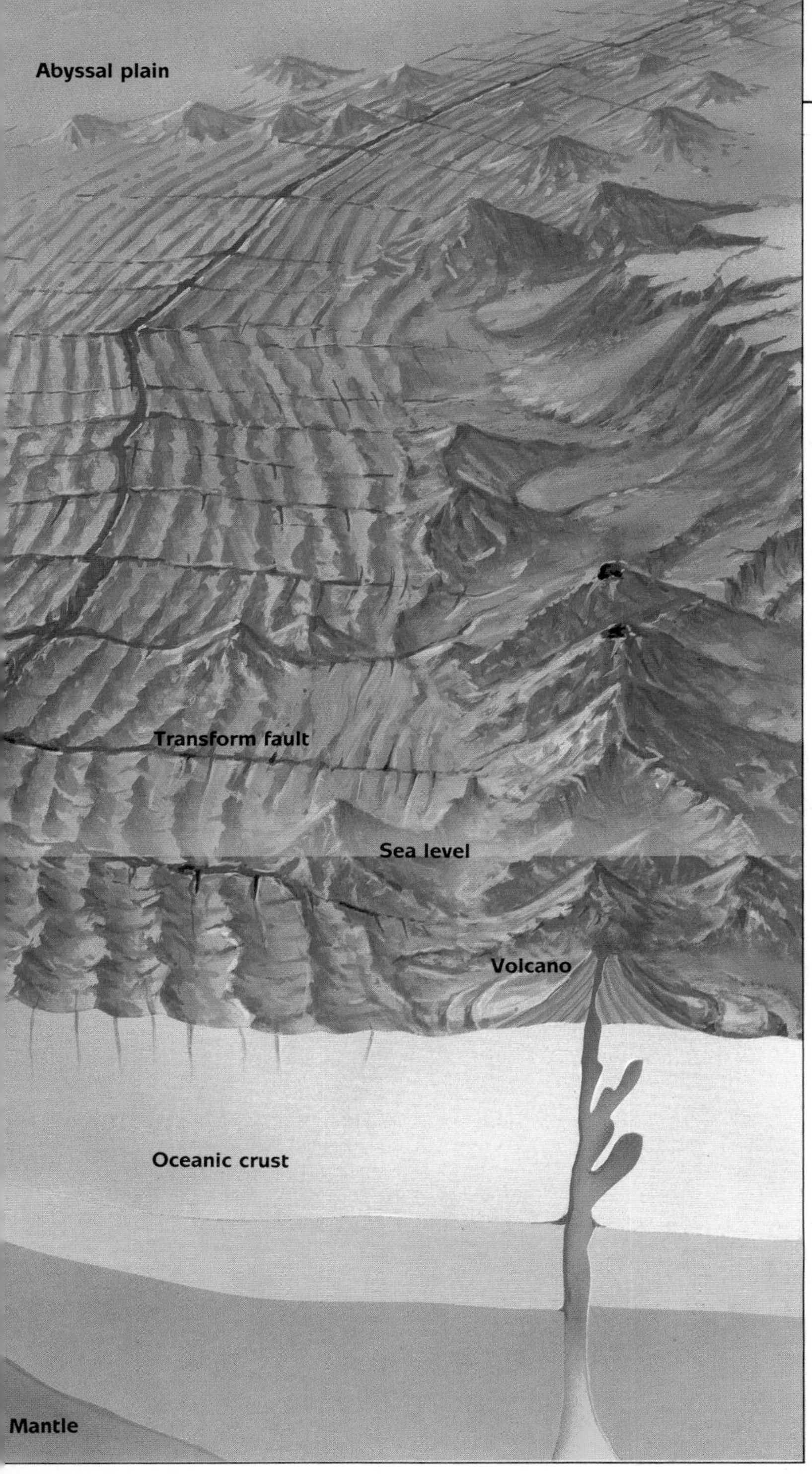


## The sinking ocean floor

At the spreading ridge the magma solidifies to form oceanic crust. Over millions of years this oceanic crust and the underlying mantle continues to cool; as the rock cools it decreases in volume causing the top of the oceanic crust to sink.

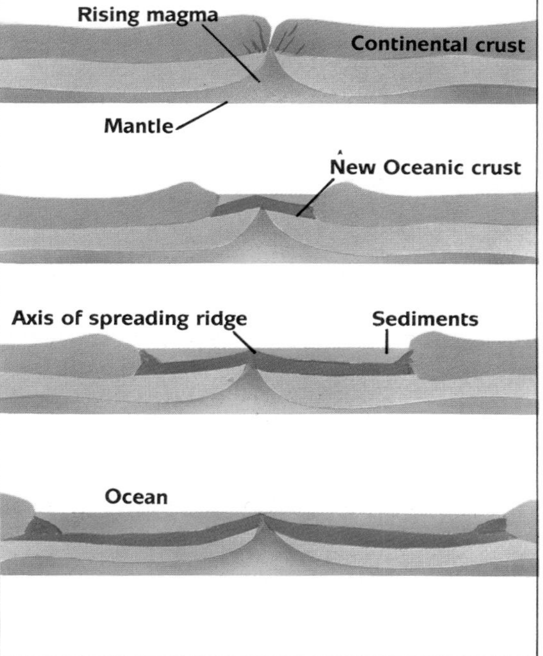


## Life on the spreading ridges

In 1977 an extraordinary discovery was made of great concentrations of living organisms growing around hot springs that emerge from the spreading ridge close to the Galapagos Islands in the eastern Pacific. Water heated by the volcanic activity at the spreading ridge emerges on the sea floor as hot springs, drawing more oceanic water to take its place. This hot water is filled with volcanic hydrogen sulfide. Around these deep hot springs, which exist in total darkness, there are thick mats of bacteria that obtain their energy through oxidizing hydrogen sulfide.

A whole range of animals cluster around the vents to browse on the bacteria, including crabs and massive clusters of clams about a foot long. Most spectacular of all are the giant worms that live in crowded colonies, each worm encased in a hard white tube, up to 10 ft (3 m) long. These worms do not need to search for food but survive by filtering the sulfide-rich sea-water and feeding off the bacteria within their own bodies.

## The oceanic food chain

The open oceans have been called the greatest deserts on Earth. Generally, the water is extremely poor in nutrients and can only support a very small number of plants, chiefly tiny phytoplankton. Grazing on the phytoplankton there are the zooplankton, including shrimps and jellyfish. Further up the food chain, there are the fast-swimming oceangoing carnivores such as the shark, dolphin, tuna and marlin, all of which have to swim great distances to find their food.

# Oceanic islands

Unique species often evolve on islands, particularly when far from the nearest landmass. Almost all oceanic islands were once volcanoes, that broke through the waves in an eruption before cooling to form a barren platform ready for life.

First to arrive on new islands are the plants. The spores of most primitive plants are so tiny that they are carried by even a light wind. Some seeds ride the moving air with wings, or fluffy parachutes. Other seeds fix with hooks or a natural glue to the feathers of birds, or may survive for long periods in a bird's intestines. The tropical coconut palm leans out over the water, and drops its large floating seeds to be carried by ocean currents to a distant beach.

For new arrivals, completing the journey is only the first of many problems. The ability of a species to adapt to new surroundings is very important, particularly if it is a bird or animal with a limited food supply. However, if the species can adapt to the smaller environment of the island it can do very well, especially if its original predators and competitors have not themselves arrived. On remote islands, animals and plants commonly adapt to new habitats, eventually evolving a wider range of species.

### Hot-spot islands

The Hawaiian island chain is the most famous example of a "hot-spot," plumes of very hot mantle that burn a hole right through the crust to emerge at the Earth's surface as large basaltic volcanoes. To the northwest are the Emperor seamounts, the submerged remnants of the older volcanic islands in the hot-spot sequence.

Dodo

### Island species

One particular property of islands is that the largest animals and the largest plants (the trees) are the least likely to make the journey. Unlike insects or lizards, no elephant will ever be accidentally transported on a floating log! The seeds of trees are themselves almost always too large and heavy to make the crossing. As a result many smaller plants evolve into tree-like forms. On the island of St. Helena in the South Atlantic, sunflowers have developed into several entirely distinct types of tree, all between 13 and 20 ft (4 and 6 m) high.

The same process can also occur for animals. On the islands of Komodo and Flores in Indonesia there is a giant lizard, up to 10 ft (3 m) long, known as the Komodo dragon. The lizard has evolved this giant form because large carnivorous mammals have not been able to make the journey to these islands, and so it has no competition. Island species commonly lose the very means of transport that originally brought them to the island, for such skills may now be wasteful. Over the generations many island birds lose the ability to fly and instead increase in size. Consequently the flightless dodo of Mauritius became highly vulnerable when a new predator, *Homo sapiens*, was carried to the island by the winds and currents from distant lands.

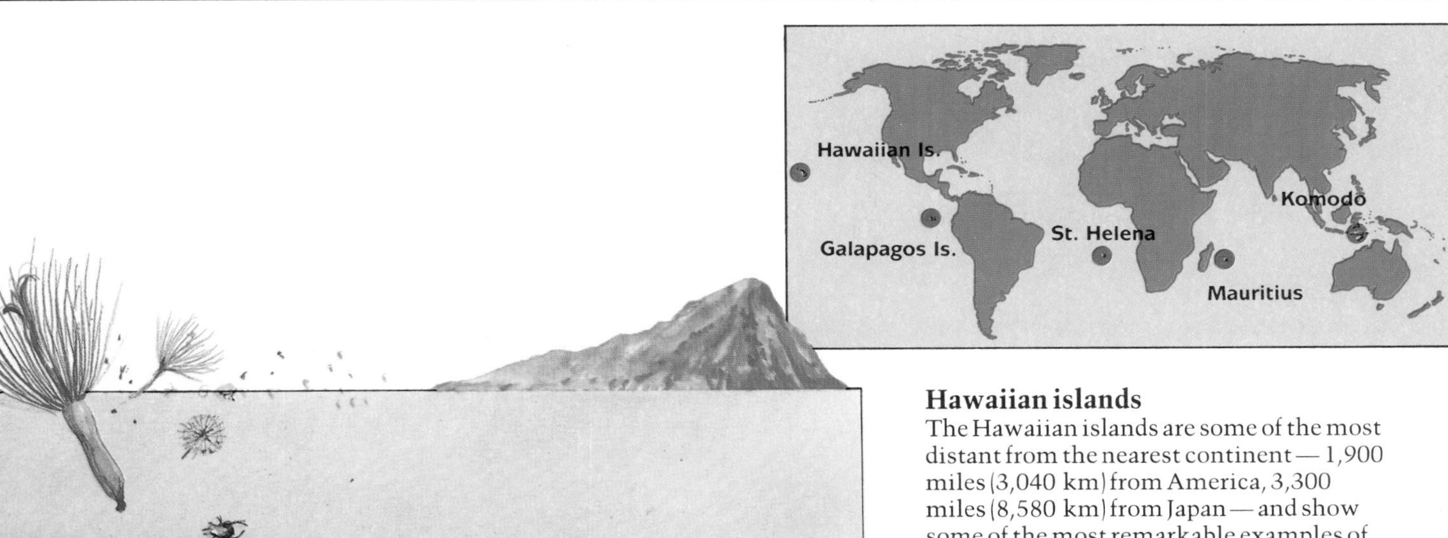

## Hawaiian islands

The Hawaiian islands are some of the most distant from the nearest continent — 1,900 miles (3,040 km) from America, 3,300 miles (8,580 km) from Japan — and show some of the most remarkable examples of isolated evolution. The islands have formed above a column of mantle, rising all the way from the Earth's core-mantle boundary. The whole Pacific plate has drifted towards the northwest over this rising "plume," which has burned a hole in the plate, leaving a trail of massive volcanic islands of basalt in its wake. New islands continue to form at the southeastern end of the chain, for example, the great dome volcanoes of Kilauea and Mauna Loa. The older islands to the northwest crumble away and sink back into the ocean, where they form the submerged Emperor chain of seamounts.

Hot-spot volcanic islands

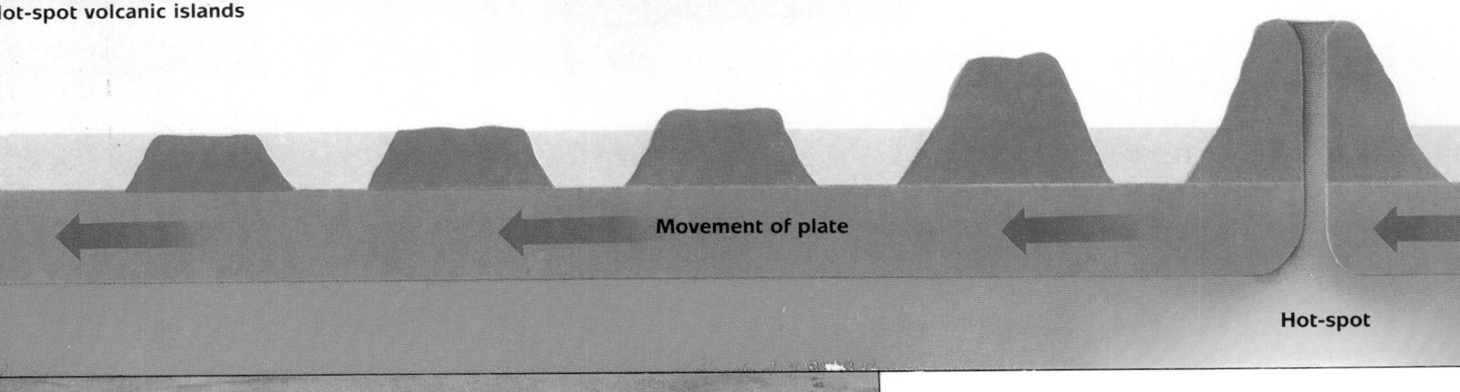


## Hawaiian animals and plants

Animals and plants have continually moved onto the new islands that form at the southeastern end of the Hawaiian chain, as the old islands to the northwest have drowned. The proportion of small-spored species such as ferns, relative to heavier-seeded trees, is far higher than on the distant continents. However one tree known as *Metrosideros* has seeds so small that it too has reached the islands. Here it has adapted to a great range of environments, being able to grow on fresh lava, in wet rain forest, as a bush on rocky mountain ridges, and even as a dwarf form in bogs.

Most of the plants arrived here from the islands close to Southeast Asia, via a series of intervening islands. However, the birds seem all to have come direct from America. Colonization requires a pair of birds traveling at almost the same time, encountering each other and breeding successfully. In the past few millions of years this improbable event has happened only 15 times. From these chance encounters have emerged a great range of different bird species, adapting to many diverse habitats.

Most notable on Hawaii are the familiar animals that are missing. Of all reptiles, amphibians, and mammals, only one (before man) has colonized the islands – a species of bat. Until man introduced them, there were even no ants on Hawaii.

From an original pair of emigrants many separate species of Hawaiian honeycreepers have evolved, adapted to different diets.

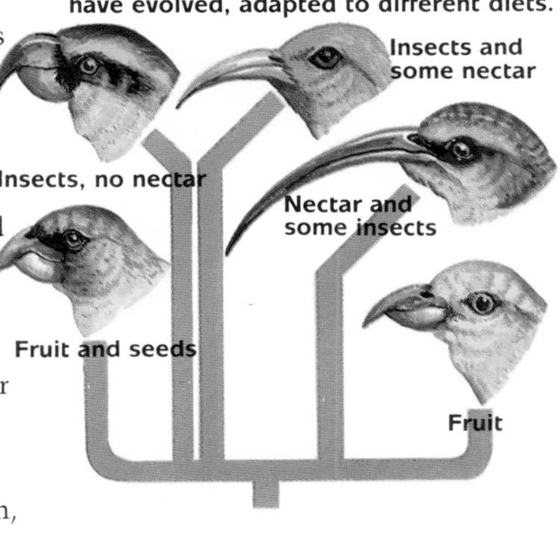

# Subduction zones

When the oceanic crust and underlying mantle (together comprising the oceanic plate) have cooled and contracted for around 50 million years, they become denser than the underlying hot mantle. The underlying rock is far too solid to allow the crust simply to founder into the mantle. Instead, the oceanic plate slides sideways at an angle down into the mantle. This process is known as subduction and the location where it occurs, the subduction zone.

As the ocean floor curves downward it creates a topographic depression (the deep ocean trench) that can be up to 7 miles (11.2 km) below sea level. The lowest point on the Earth's surface lies in the Marianas Trench in the Pacific Ocean to the south of Japan. Ahead of the trench, sediments are scraped off the sinking ocean floor. Some islands, such as Barbados in the West Indies, or Timor in Indonesia, have formed where this pile of sediment scrapings rises above the surface of the sea.

At a depth of 60 to 90 miles (100–150 km) the ocean plate melts, producing magma, which rises to the surface as volcanoes. The pattern of a deep ocean trench running parallel to a chain of volcanoes, is the signature of an underlying subduction zone.

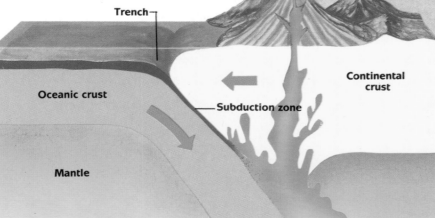

Trench
Oceanic crust
Subduction zone
Continental crust
Mantle

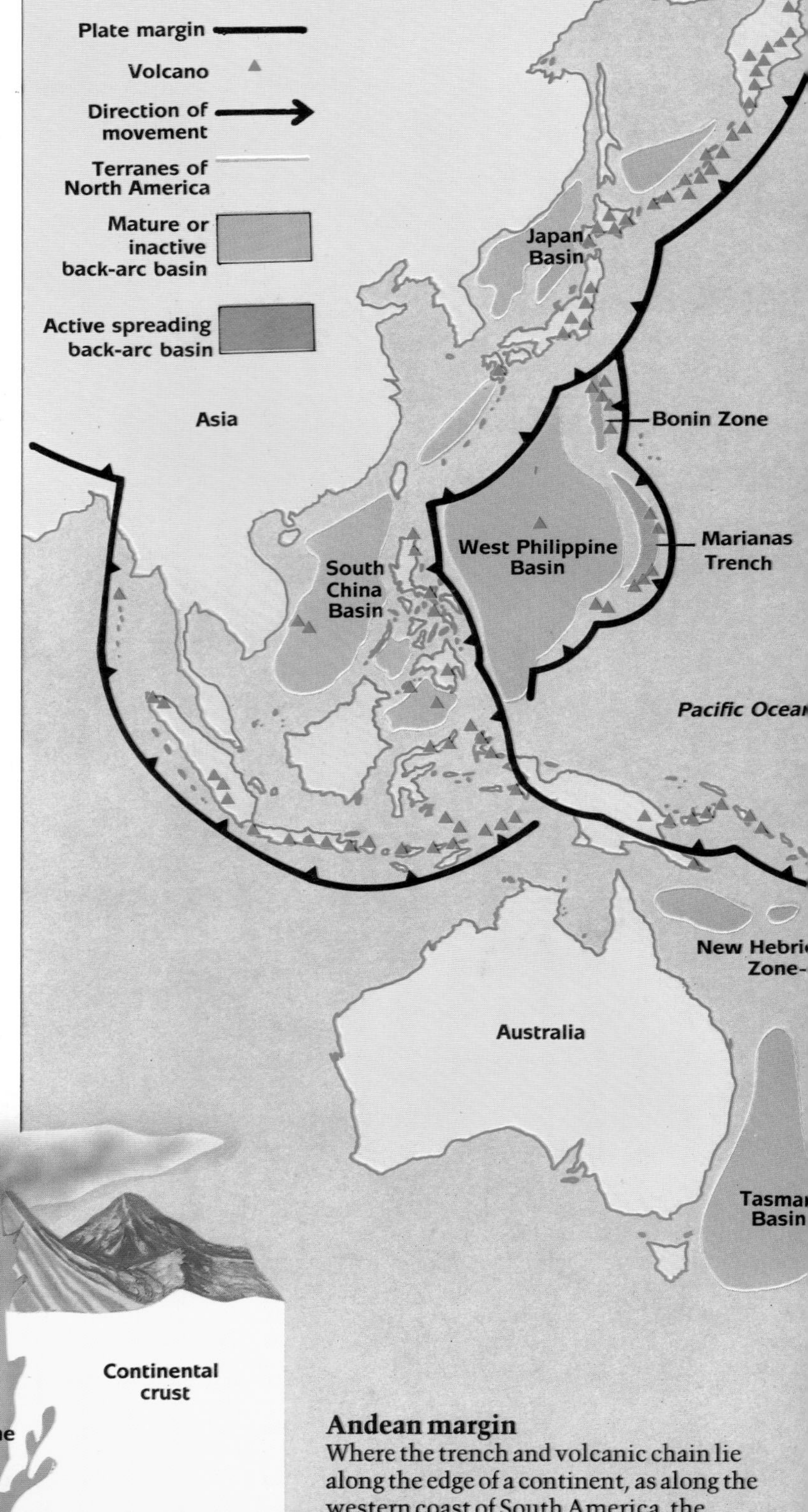


**Andean margin**
Where the trench and volcanic chain lie along the edge of a continent, as along the western coast of South America, the subduction zone forms what is called an "Andean margin." Where the volcanic chain is a garland of islands, it is called an "island arc." Many island arcs started life as Andean margins, but became separated from the mainland.

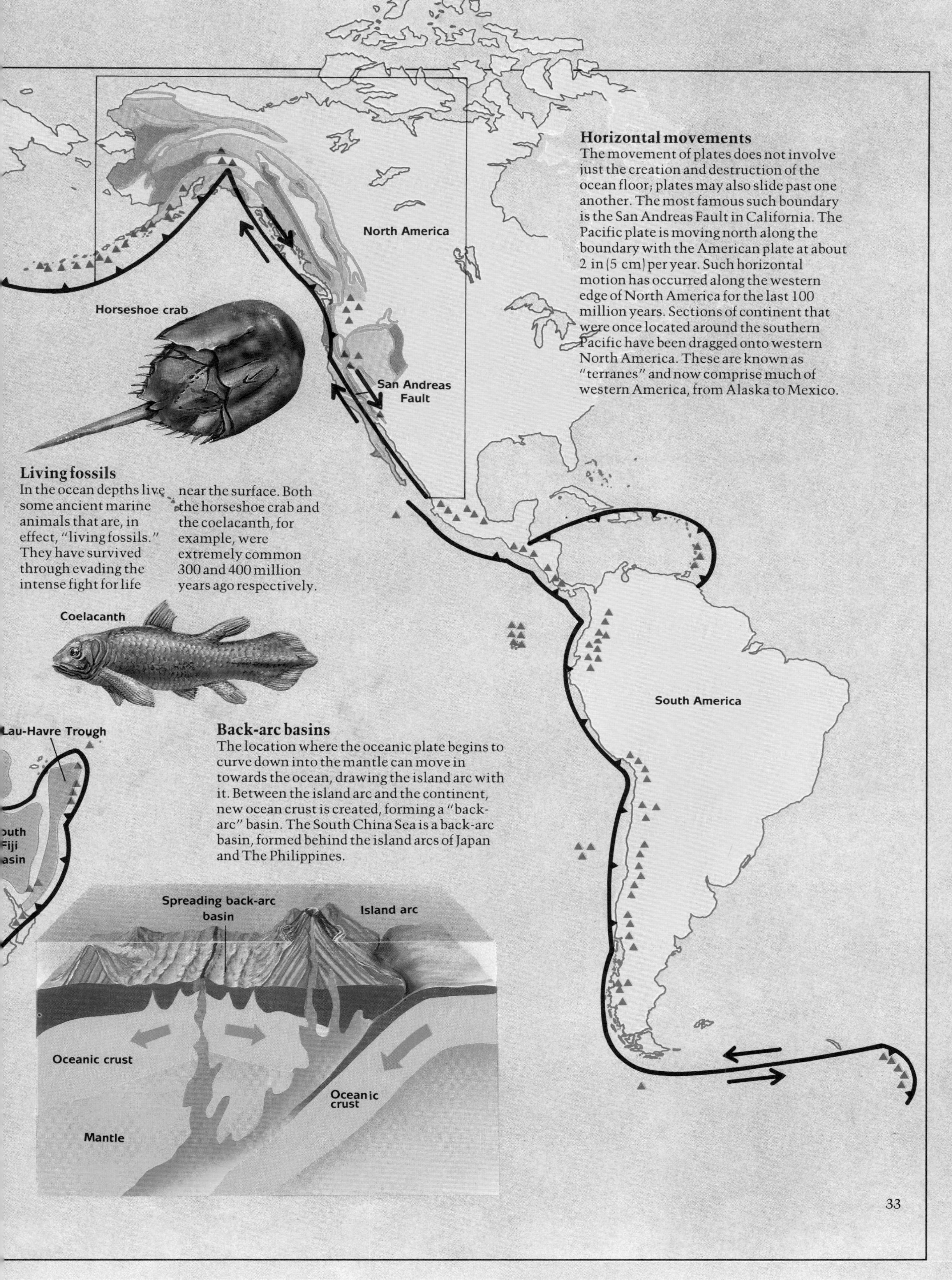


## Horizontal movements

The movement of plates does not involve just the creation and destruction of the ocean floor; plates may also slide past one another. The most famous such boundary is the San Andreas Fault in California. The Pacific plate is moving north along the boundary with the American plate at about 2 in (5 cm) per year. Such horizontal motion has occurred along the western edge of North America for the last 100 million years. Sections of continent that were once located around the southern Pacific have been dragged onto western North America. These are known as "terranes" and now comprise much of western America, from Alaska to Mexico.

## Living fossils

In the ocean depths live some ancient marine animals that are, in effect, "living fossils." They have survived through evading the intense fight for life near the surface. Both the horseshoe crab and the coelacanth, for example, were extremely common 300 and 400 million years ago respectively.

## Back-arc basins

The location where the oceanic plate begins to curve down into the mantle can move in towards the ocean, drawing the island arc with it. Between the island arc and the continent, new ocean crust is created, forming a "back-arc" basin. The South China Sea is a back-arc basin, formed behind the island arcs of Japan and The Philippines.

EVOLVING ENVIRONMENTS

# Colliding continents

After 200 million years, the original fragments of the Gondwana continent had moved so far apart that it was inevitable they should once again come into contact with the northern supercontinent, Laurasia, which itself had broken into two massive pieces – North America and Eurasia.

Of the five major fragments of Gondwana, only Antarctica had no contact with Laurasia, and if its wildlife had not been exterminated by its movement into the highest southern latitudes, it would preserve the purest Gondwana fauna and flora of all. However, both South America and Australia maintained their independence until relatively recently, preserving much of their Gondwanan wildlife. We can observe the encounter between Laurasia and Gondwana, at different stages of the contest: in Indonesia and Panama.

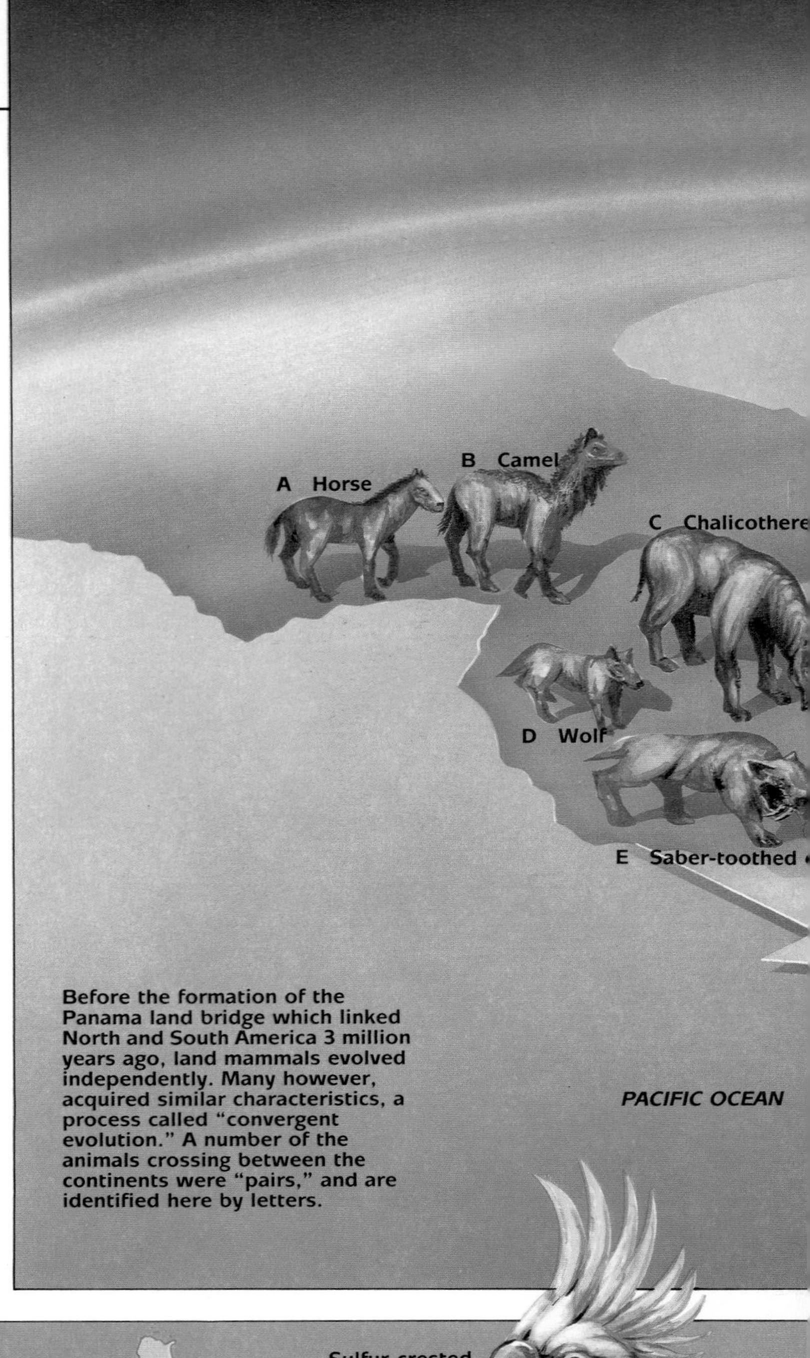


Before the formation of the Panama land bridge which linked North and South America 3 million years ago, land mammals evolved independently. Many however, acquired similar characteristics, a process called "convergent evolution." A number of the animals crossing between the continents were "pairs," and are identified here by letters.

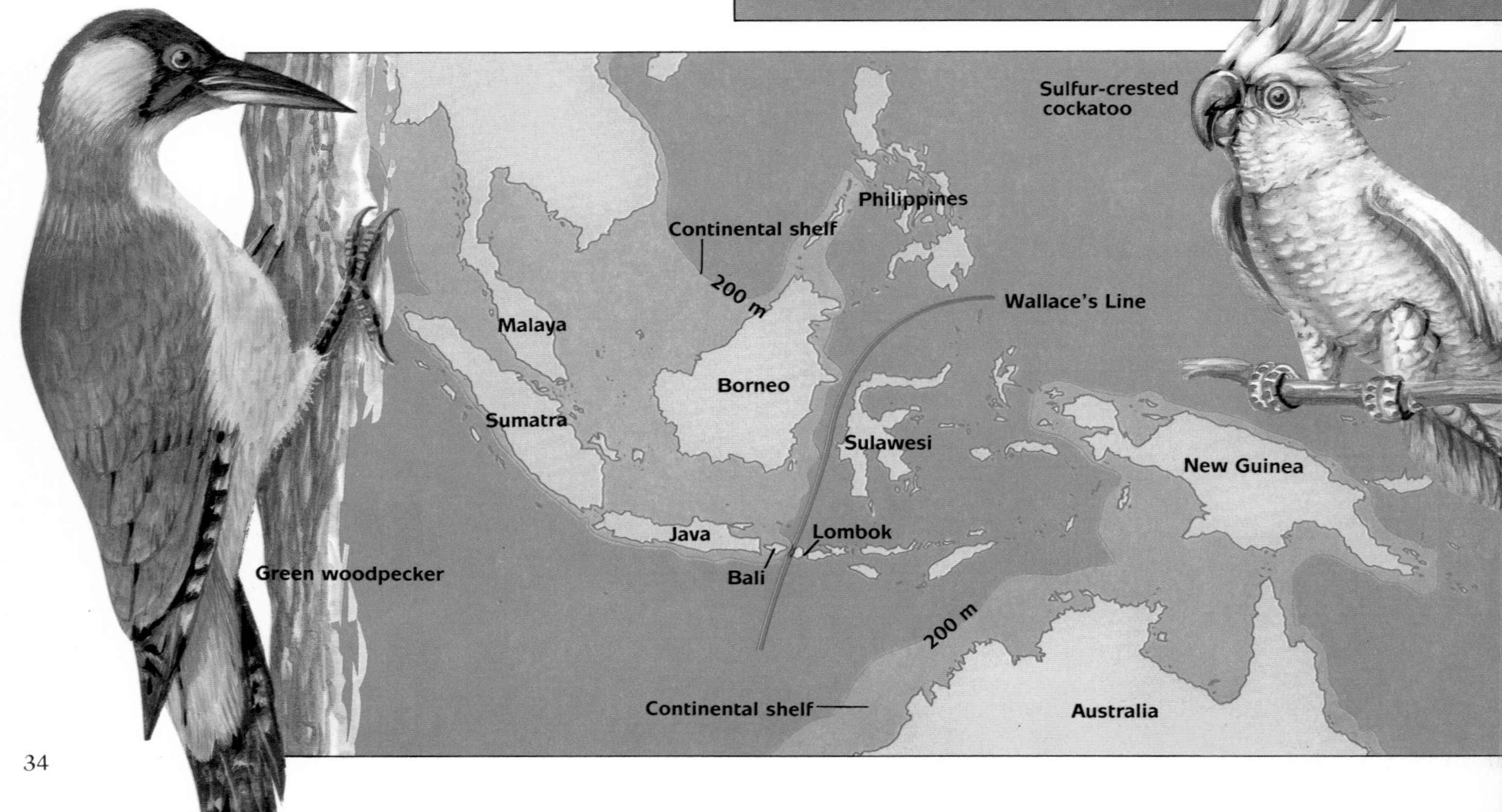

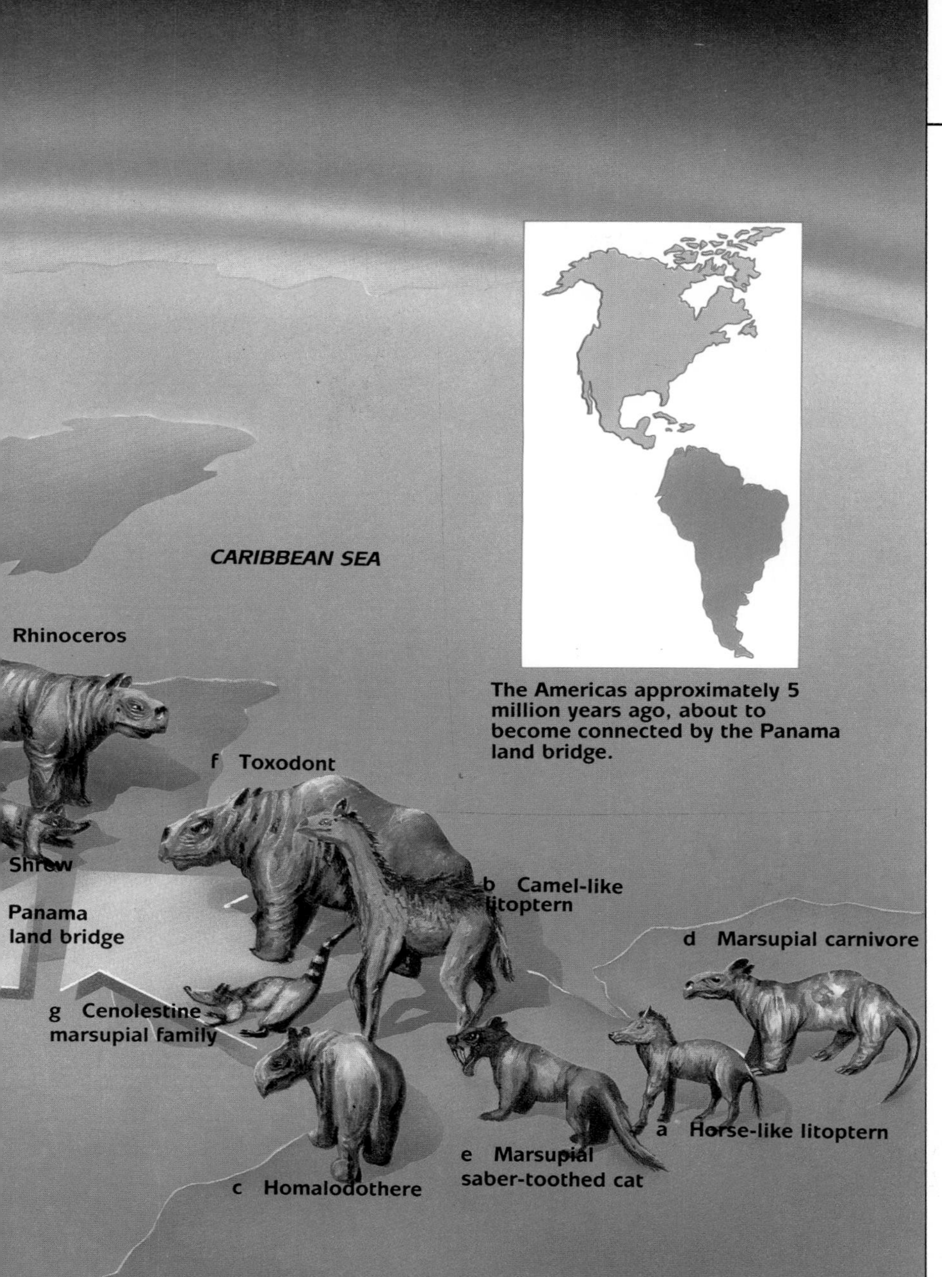


The Americas approximately 5 million years ago, about to become connected by the Panama land bridge.

## North versus South America

South America never preserved a pure Gondwana fauna. At several times in the past 100 million years there was partial contact with other continents. About 35 million years ago, for example, some monkeys and rodents arrived from Africa, perhaps via a chain of mid-Atlantic volcanic islands, connecting west Africa with Brazil.

Around 10 million years ago a chain of islands connected North and South America. By 3 million years ago a land bridge through Panama had become a permanent link. Both continents had their own land mammals and many of these crossed into the other continent. Those traveling north, however, found their way blocked by the Mexican desert, while North American species already resident in Central America moved unimpeded into the Amazon forests. The contest soon became very one-sided.

The dramatic climatic changes of the Ice Age caused a new wave of extinctions in South America – giant armadillo-like glyptodonts and ground sloths died out, as well as horses and elephants that had come from the north. Of the tropical animals from the South, only the opossum, armadillo and anteater have survived as residents in the north. The main losers in South America were the marsupials, who were simply unable to compete with the northern mammals. Today only a few species remain.

During exchange
Northern species 62%
Southern species 38%
Northern species 35%
Southern species 65%

After exchange
Northern species 82%
Southern species 18%
Northern species 41%
Southern species 59%

After the Panama land bridge had linked the two Americas, northern land mammal species proved more successful in the south than southern ones in the north.

## Indonesia

While the Indonesian chain of islands has linked the two continents for the past 10 million years, at no time has there been a land bridge connecting Australia with Asia, even when sea levels were low during the Ice Age.

Defining precisely where Australia starts and Asia stops depends on which organisms are being studied. If the boundary were to be defined from the spread of mammals, then the encounter has not yet begun, for the intervening islands, which lie between the Australian and Asian continental shelves, have very few indigenous mammals of either origin. In terms of flowering tropical plants, Asia extends as far east as New Guinea, while for birds the boundary has to be drawn farther to the west, between the tiny neighboring islands of Bali and Lombok. In terms of insects New Guinea is Asian, although the fact that marsupial mammals have reached this large island suggest rather that it is Australian.

For other species the very idea of a boundary is too simplistic. The Australian eucalyptus and casuarina trees have already advanced as far as Malaysia. Around 2 million years ago the collision between the continents created a series of high mountain ranges on the individual islands that were close enough to one another to create a chain of cool habitats, within settling distance of windblown seeds. Some temperate plants (veronica and euphrasia) have crossed from the northern to the southern hemisphere, high above the hot tropical lowlands.

Apart from the bats, the only group of placental mammals that reached Australia were the rats, which now number 50 species and comprise half of Australia's land mammal species. The Australian dingo is descended from domestic dogs, introduced only 3,500 years ago.

EVOLVING ENVIRONMENTS

# Mediterranean Sea

The Mediterranean is a unique sea, formed in the middle of a huge collision between Africa and Eurasia. These continents were originally separated by an ocean named "Tethys," an ocean that broadened and divided into two separate arms to the east.

About 50 million years ago Spain collided with France, throwing up the Pyrenean mountains that formerly continued across southern France. To the southeast of France lay the western end of Tethys. Soon a subduction zone developed at the northern edge of the ocean, consuming the ocean crust. The eastern half of the mountain chain began to stretch and collapse, the crust becoming so thin that the underlying mantle melted. Fragments of this thin continent became separated from France, and today form the islands of Sardinia, Corsica, and the Balearics.

The subduction zone continued to shift southwards, and new ocean crust began to develop to the south of these islands in the Tyrrhenian Sea. Finally the subduction zone crashed into the northern coast of Africa, throwing up a chain of mountains through Algeria.

35 million years ago

Corsica

Sardinia

Sicily

Present-day coastlines are superimposed on maps of shifting plate margins.

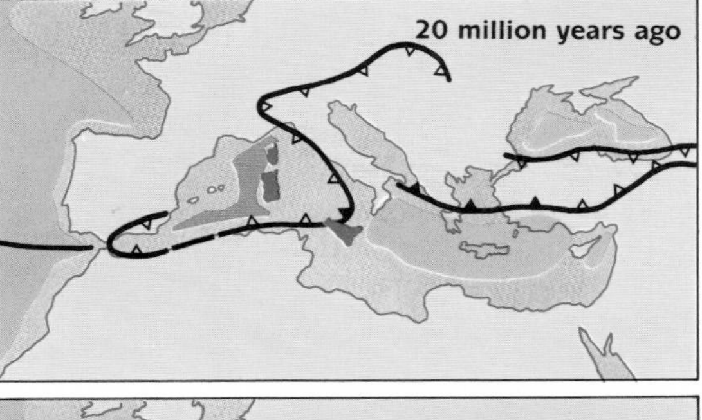


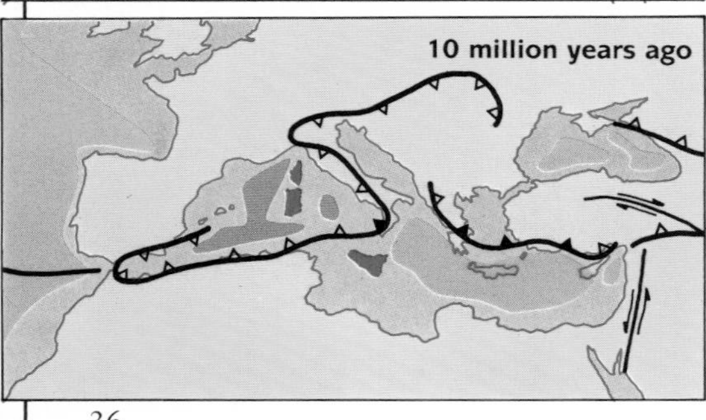


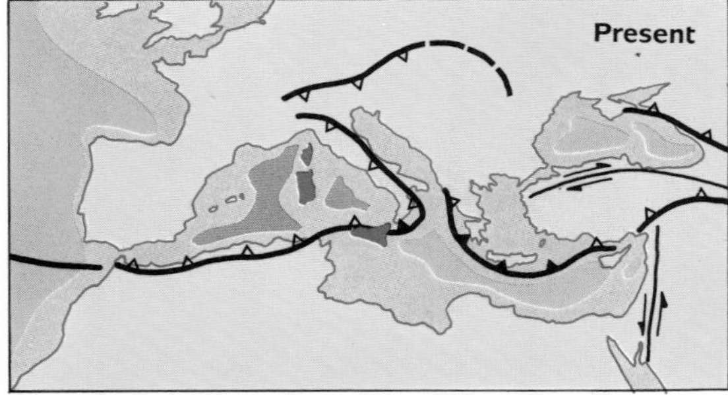


**The Mediterranean desert**
As a landlocked sea in a warm climate, the Mediterranean loses around 800 cubic miles of water each year through evaporation. This is about three times as much as is replenished by rain and rivers, and an enormous river of Atlantic water flows through the Strait of Gibraltar to keep the sea filled. Yet 6.5 million years ago, this Atlantic connection became closed. In the space of about 1,000 years the whole Mediterranean Sea shrunk to a few central salt lakes. The whole sea floor, 1 to 2 miles (1.6 to 3.2 kilometers) deep, lay exposed to view. Then, perhaps as a result of a major earthquake in southern Spain, or a global rise in sea level, the Atlantic once again started to flow into the

Strike-slip faul
Collision zone
Subduction zo
Plate boundar

Mediterranean, in a vast cascade, with a flow thought to be 10,000 times greater than Niagara Falls. In time this flow refilled the basin, before the connection became blocked once again.

When the Mediterranean Sea dried out, it became possible to walk to today's islands, such as Cyprus and Corsica. It was probably at this period that a variety of mammals colonized the islands, including ancient types of rabbit and antelope.

During the last Ice Age, possibly at a time of reduced sea levels, a number of Mediterranean islands were colonized by hippos and elephants. As food was relatively scarce the hippos and elephants had to evolve smaller forms to survive. On the other hand, dormice on Sicily and Mallorca exploited the absence of carnivores and evolved into giant dormice, as large as rabbits. This strange mixture of dwarf and giant mammals survived through the last Ice Age up until the arrival of man.

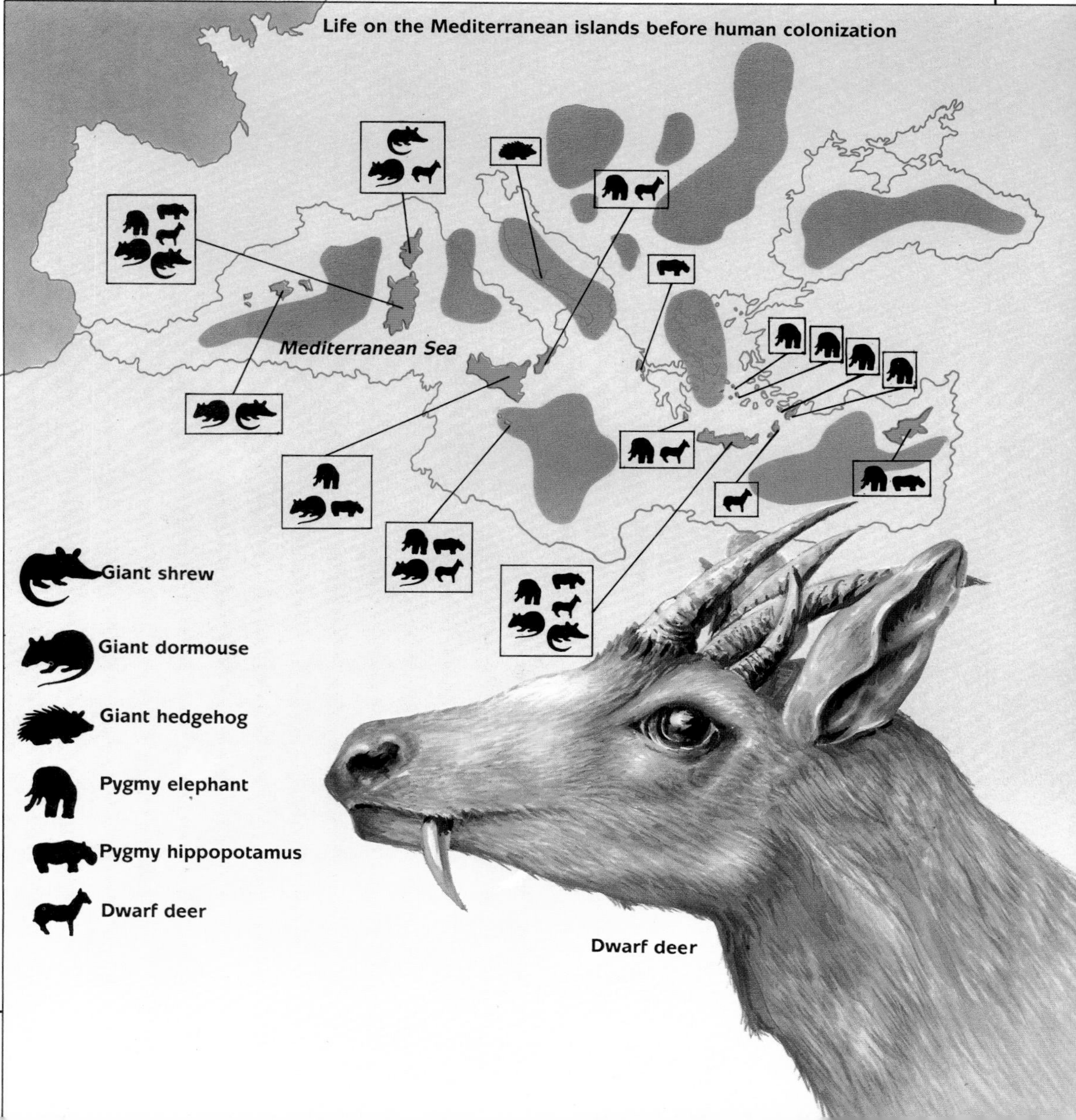

EVOLVING ENVIRONMENTS

# The Himalayas

Great mountain ranges, or massifs, are built as continents collide. The greatest of all today's massifs is the Himalayan-Karakoram Range. This is one section of a complex band of mountains, known as the "Alpine Chain" that runs from the Pyrenees through to Indonesia. Some of this mountain chain formed 50 million years ago, and is now rounded and mature; other, jagged summits continue to grow.

The Alpine Chain represents the collision of continents when the former ocean of Tethys became swallowed up by the mantle. In the midst of this titanic collision the crust buckled at its weakest points, each of these buckles generating a mountain range. The collision is complicated because there was not simply an Asian and an Indian continent, but a pattern of lesser islands within Tethys, which have now all become compressed together.

The Himalayas lie along a gentle curve between the Tibetan plateau and India. However, to the northwest there is a great knot of mountains, with the Karakoram at its center. This remarkable range contains a mountain (K2) that rises to 28,500 ft (8,500 m) second only in elevation to Mt Everest.

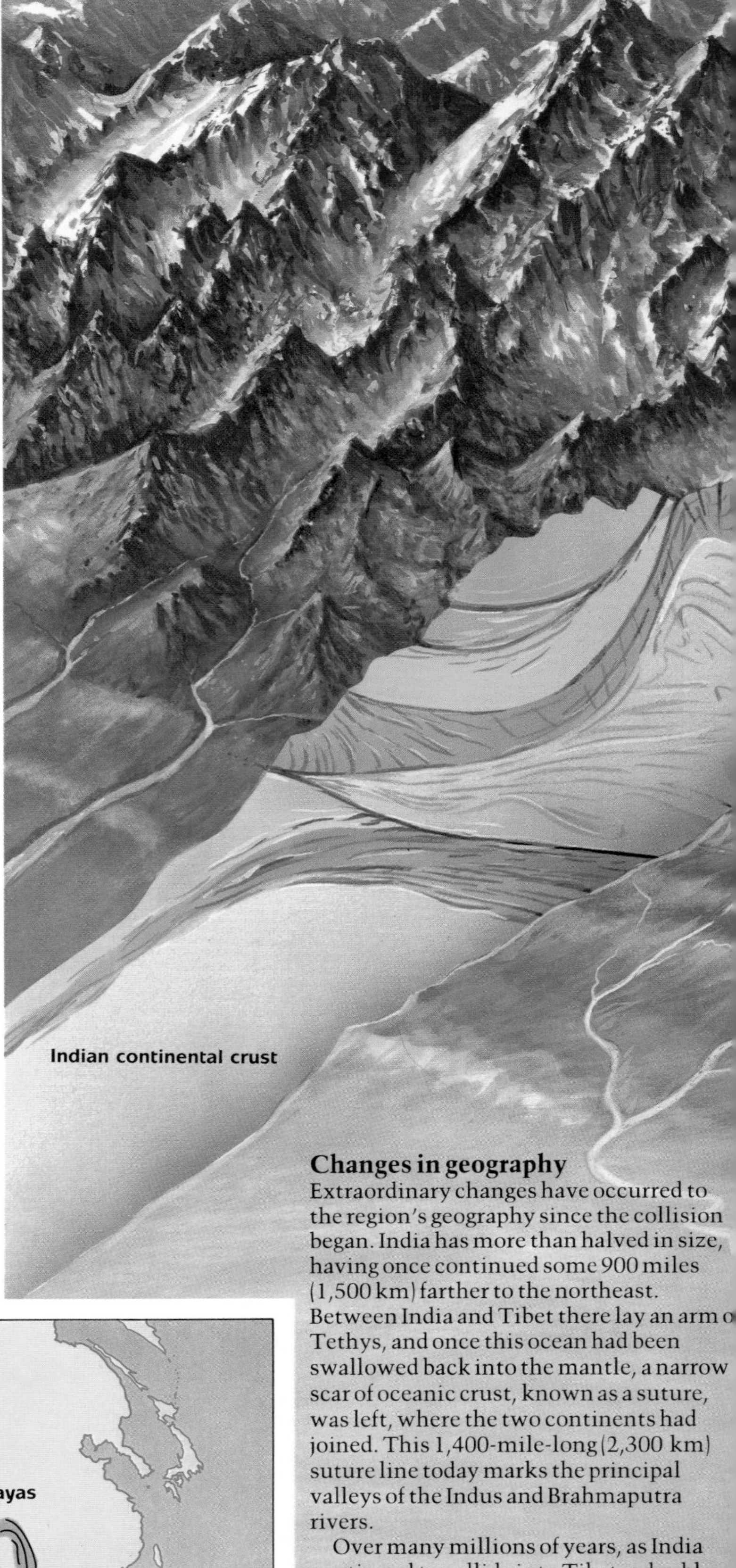


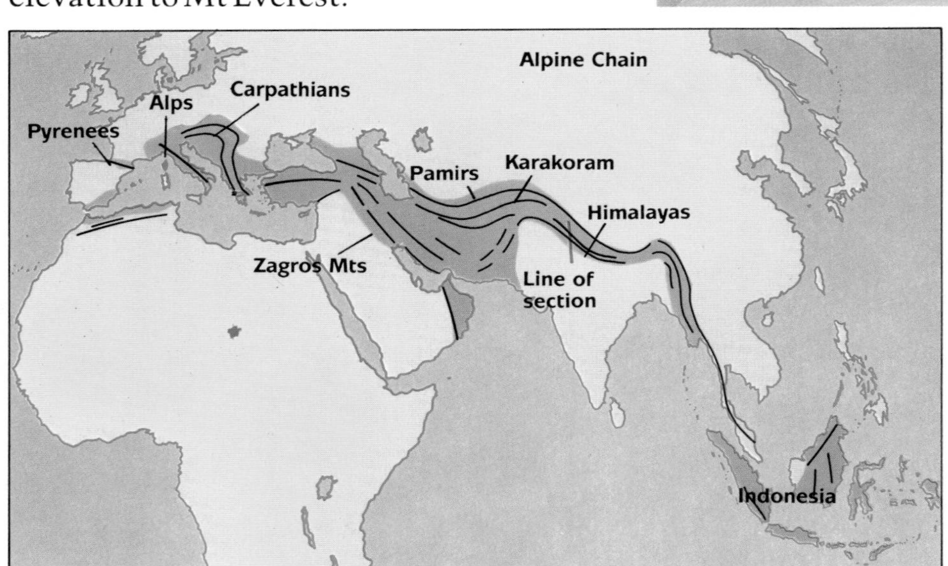


### Changes in geography

Extraordinary changes have occurred to the region's geography since the collision began. India has more than halved in size, having once continued some 900 miles (1,500 km) farther to the northeast. Between India and Tibet there lay an arm o Tethys, and once this ocean had been swallowed back into the mantle, a narrow scar of oceanic crust, known as a suture, was left, where the two continents had joined. This 1,400-mile-long (2,300 km) suture line today marks the principal valleys of the Indus and Brahmaputra rivers.

Over many millions of years, as India continued to collide into Tibet, a double-thick 40-mile (65 km) crust was created beneath Tibet. The main concentration of the collision is along the southern edge of the Tibetan plateau, where the crust has crumpled and been thrust high into the sky. The highest point on Earth, Mt Everest, is composed of limestone laid down in a shallow sea on the edge of a grea Tethyan island almost 300 million years ago.

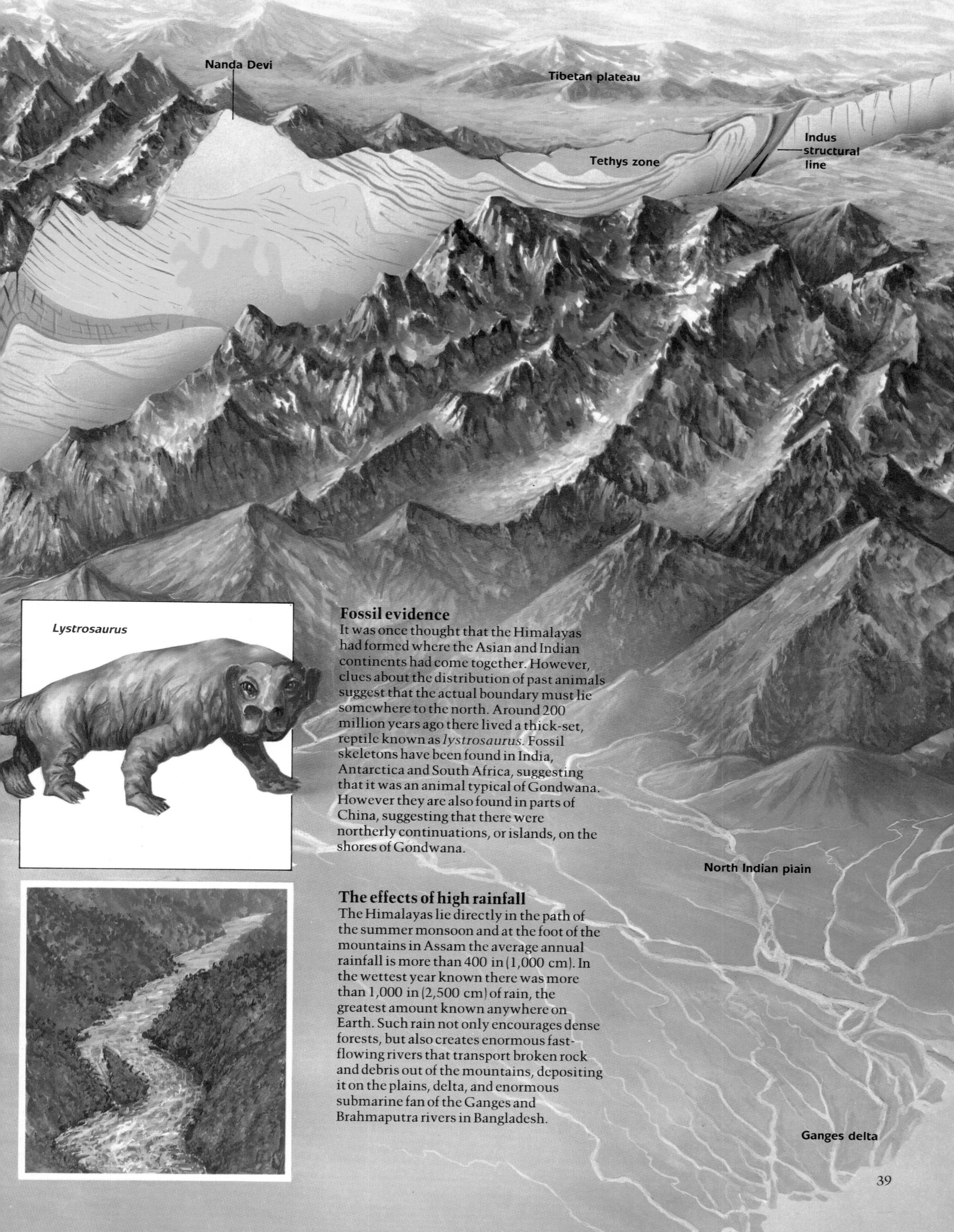


## Fossil evidence

It was once thought that the Himalayas had formed where the Asian and Indian continents had come together. However, clues about the distribution of past animals suggest that the actual boundary must lie somewhere to the north. Around 200 million years ago there lived a thick-set, reptile known as *lystrosaurus*. Fossil skeletons have been found in India, Antarctica and South Africa, suggesting that it was an animal typical of Gondwana. However they are also found in parts of China, suggesting that there were northerly continuations, or islands, on the shores of Gondwana.

## The effects of high rainfall

The Himalayas lie directly in the path of the summer monsoon and at the foot of the mountains in Assam the average annual rainfall is more than 400 in (1,000 cm). In the wettest year known there was more than 1,000 in (2,500 cm) of rain, the greatest amount known anywhere on Earth. Such rain not only encourages dense forests, but also creates enormous fast-flowing rivers that transport broken rock and debris out of the mountains, depositing it on the plains, delta, and enormous submarine fan of the Ganges and Brahmaputra rivers in Bangladesh.

# The cold shield

Even great mountain ranges fall victim to erosion. Glaciers, rockfalls, and torrents – all remove rubble from the mountains.

Four-fifths of the rocks of a mountain range lie buried, floating on the mantle. After perhaps 100 million years, the exposed rocks are those that were buried several miles beneath high mountain peaks. At such depths in the crust, rocks are hot, even molten. While they might once have been sediments laid down in a lake or sea, heat and pressure deforms and reconstitutes them as "metamorphic" rocks.

Much of the crust of the continents has been created in past mountain ranges and so the rocks are metamorphic, sometimes exposed at the surface or covered by thick soils or sediments. These are the old continental shields of Africa, Russia, and the Americas, forged in collisions between continents over one billion years ago.

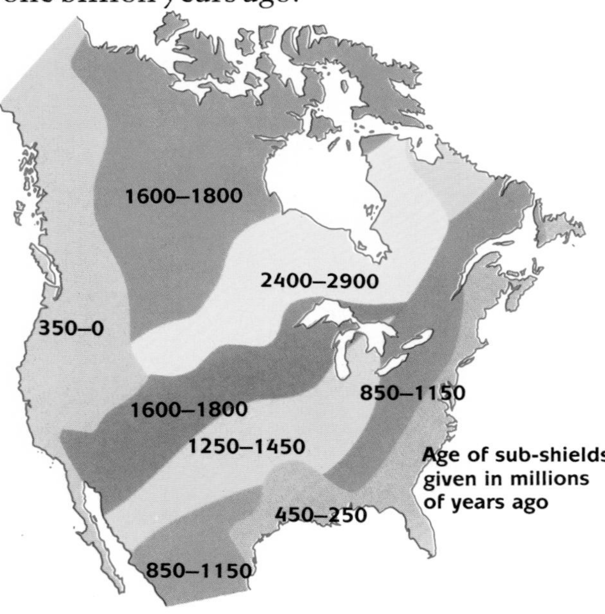


**North American sub-shields**
The North American continent has grown like a snowball over billions of years; a series of continental shields wrap around each other, becoming younger towards the continental margins. Each shield is the eroded remains of mountain ranges formed as continents collided.

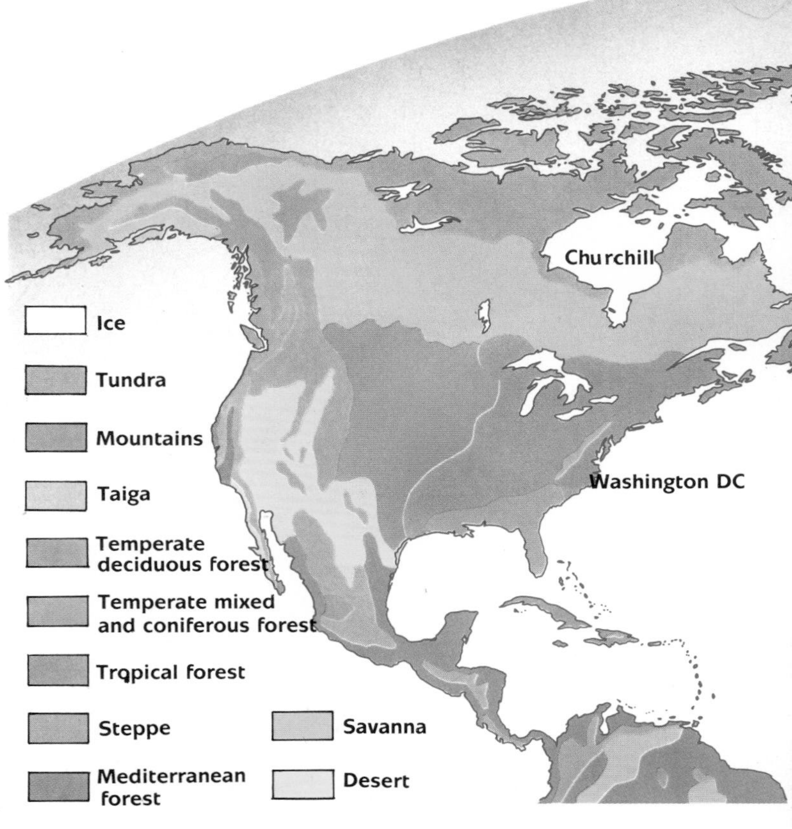


**Biomes**
Plants are good climatic indicators because the conditions under which seeds germinate, plants grow, and new seeds form, are often very sensitive. Different climatic zones can be identified from their characteristic vegetation; these localized groups of plants are known as "biomes." In the Arctic, plants grow low, to escape the fierce winds and abrasive ice crystals. In temperate regions, protection against winter winds is achieved by concentrating growth in the summer and retreating into dormancy in the winter – a property of the deciduous trees.

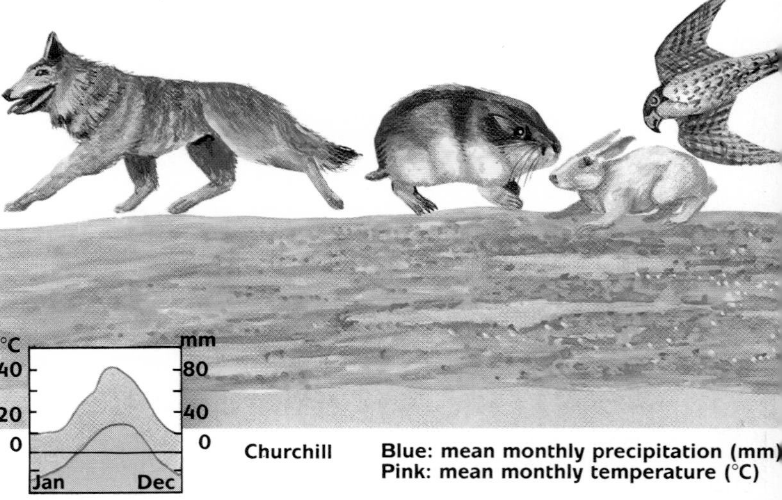


**Tundra**
A permanent layer of frozen soil (permafrost) persists even in the heat of midsummer. The only plants living here are gnarled dwarf trees, sedges, lichens, and mosses. These all grow in a short summer when they become food for reindeer and musk-ox, snow hares and lemmings. Wolves, owls, hawks, and foxes prey on the herbivores. Many birds migrate to the tundra in summer to feast on the millions of insects that breed in the bogs of melted permafrost.

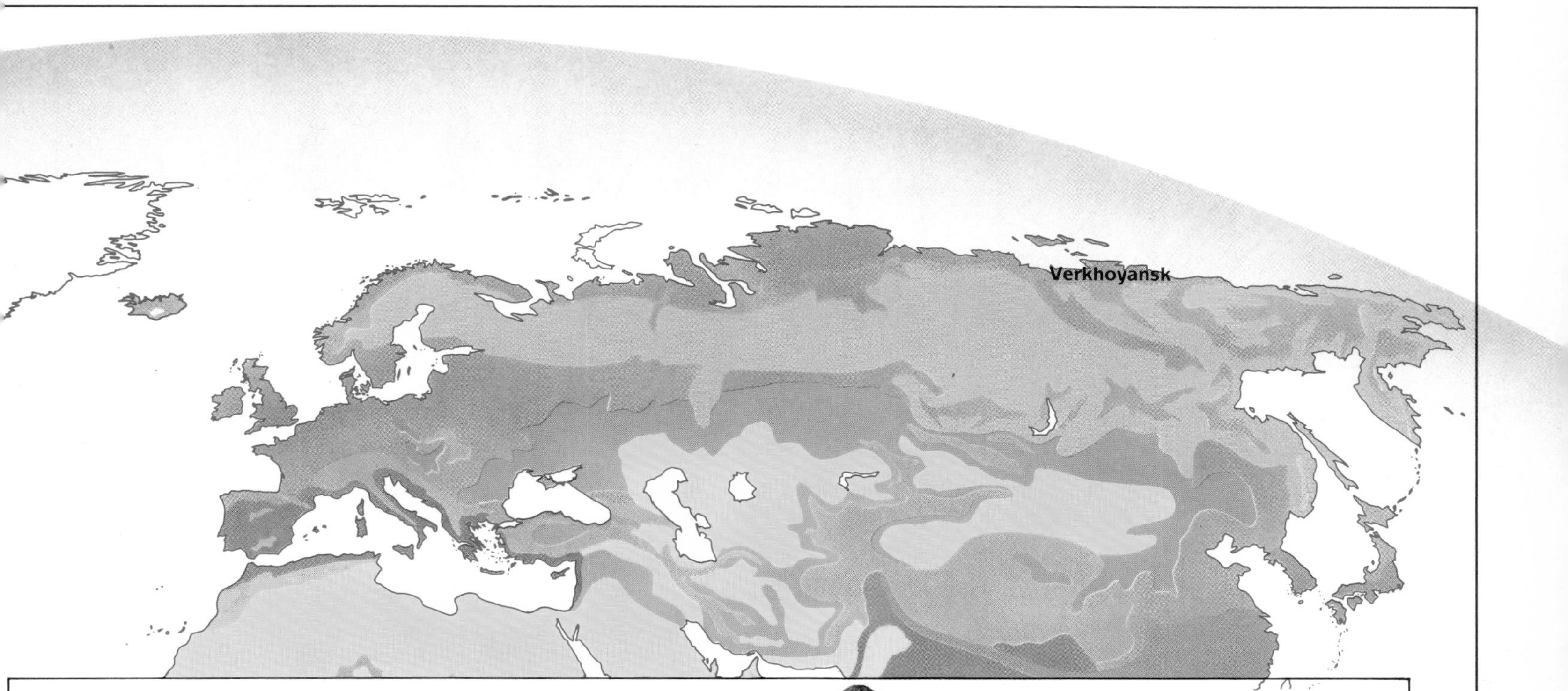

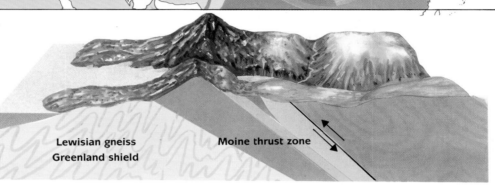

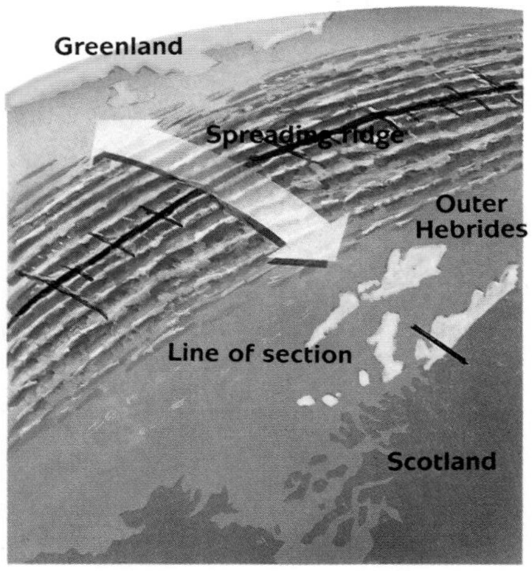

**Broken shield**
On rare occasions old shields become broken, the two parts separated by new oceanic crust. At the far northwestern edge of Britain, the Outer Hebrides islands form part of a continental shield, almost 3 billion years old, that was fractured off the eastern edge of Greenland only 55 million years ago.

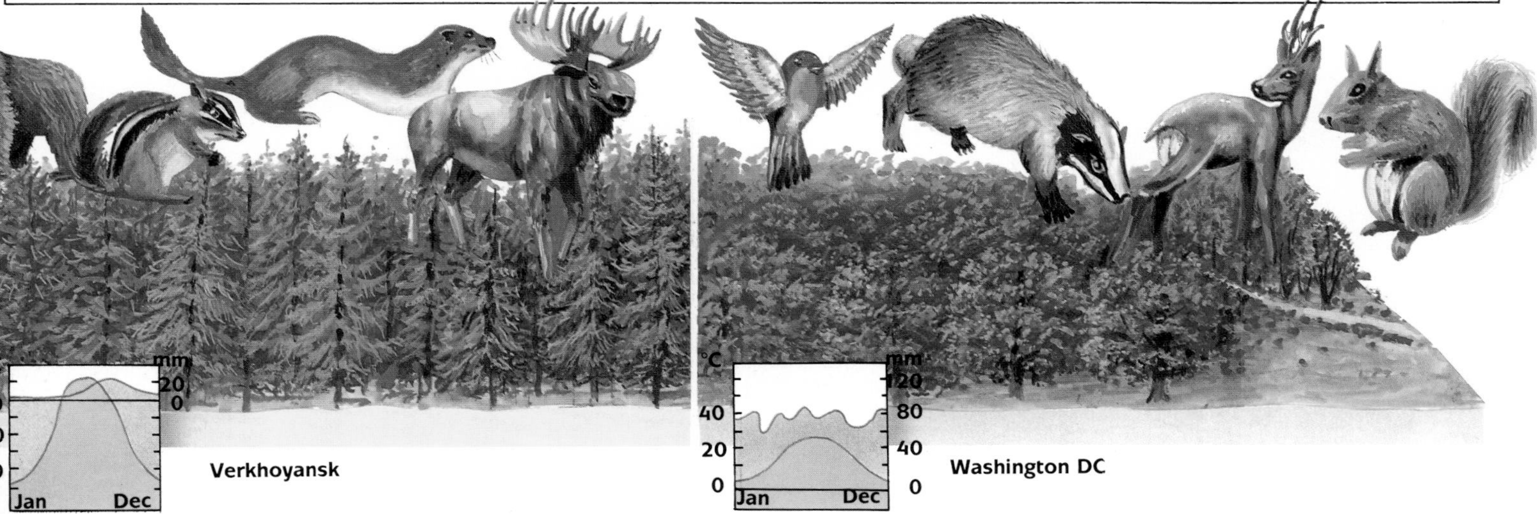


**Taiga**
Coniferous trees characterize the taiga. Employing anti-freeze sap, they can generate energy by photosynthesis even in midwinter sunshine. Their waxed needles can withstand drought, while their drooping branches resist the wind and discard a heavy fall of snow. Taiga forests are monotonous, with few species of tree, and this limits the variety of animals. Most important are the deer, and burrowing rodents that hibernate underground.

**Temperate forest**
The temperate forest includes many tree types, a blend of broadleaf deciduous trees, conifers, or broadleaved evergreens according to climate and rainfall.

Animal life includes badgers, bears, wild boar, deer, squirrels, and many rodents. On these hunt wolves, owls, wild cats and foxes.

EVOLVING ENVIRONMENTS

# The hot shield

Towards the equator the diversity of life (the number of different species of animals and plants) increases. It is the great range of flowering plants and the unseen chorus of innumerable birds that characterize the tropics, relative to the temperate woodlands of higher latitudes.

There are several reasons for this increase in the number of species. First, all the temperate regions suffered greatly during the last Ice Age. Extensive areas were turned to tundra or were covered by glaciers. Many of the species that formerly inhabited these regions were destroyed as the ice caps advanced. Animals and plants at high latitudes are almost all relatively new arrivals; as new populations, there has been no time for them to have evolved into further new species. Second, temperate regions are more likely to suffer occasional extremes of climate – a fierce winter or a prolonged drought – while conditions closer to the equator are far more constant. Finally, there are more varied habitats in the tropics: While high ground at the equator has a climate much like that of the temperate regions, there are no local tropical climates in the temperate belt.

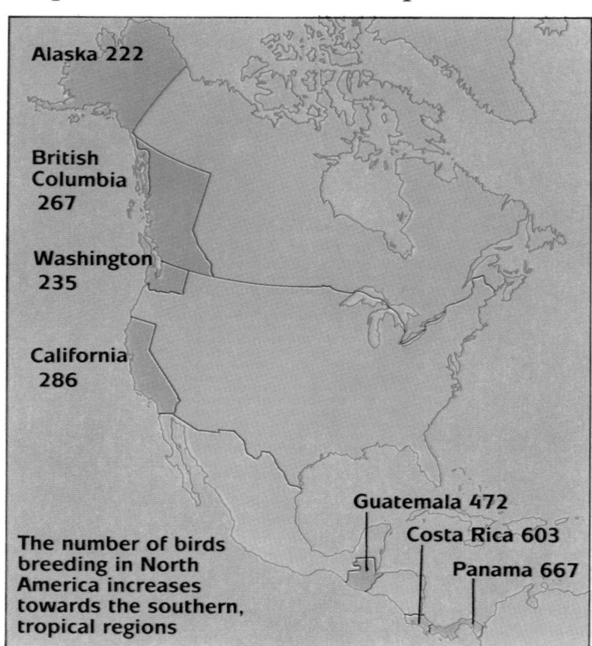

The number of birds breeding in North America increases towards the southern, tropical regions

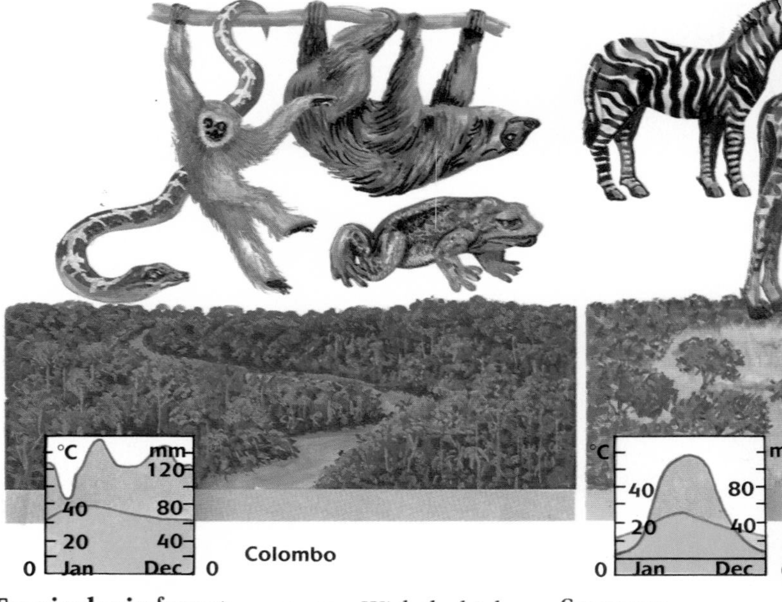


**Tropical rain forest**
The tropical rain forests can occur anywhere in the tropics, but they demand a high and regular rainfall. At ground level the forest is surprisingly dark; most sunshine is filtered out by the dense canopy. With the highest productivity of living matter of any environment on Earth, the forest is rich in species, more so, probably, than at any time in the past 300 million years.

**Savanna**
Where rainfall is insufficient for tropic forest, tall grassland vegetation rising to 12 ft (3.5 m) develops. Known as savanna, it commonly contains a scattering of drought-resistant trees and

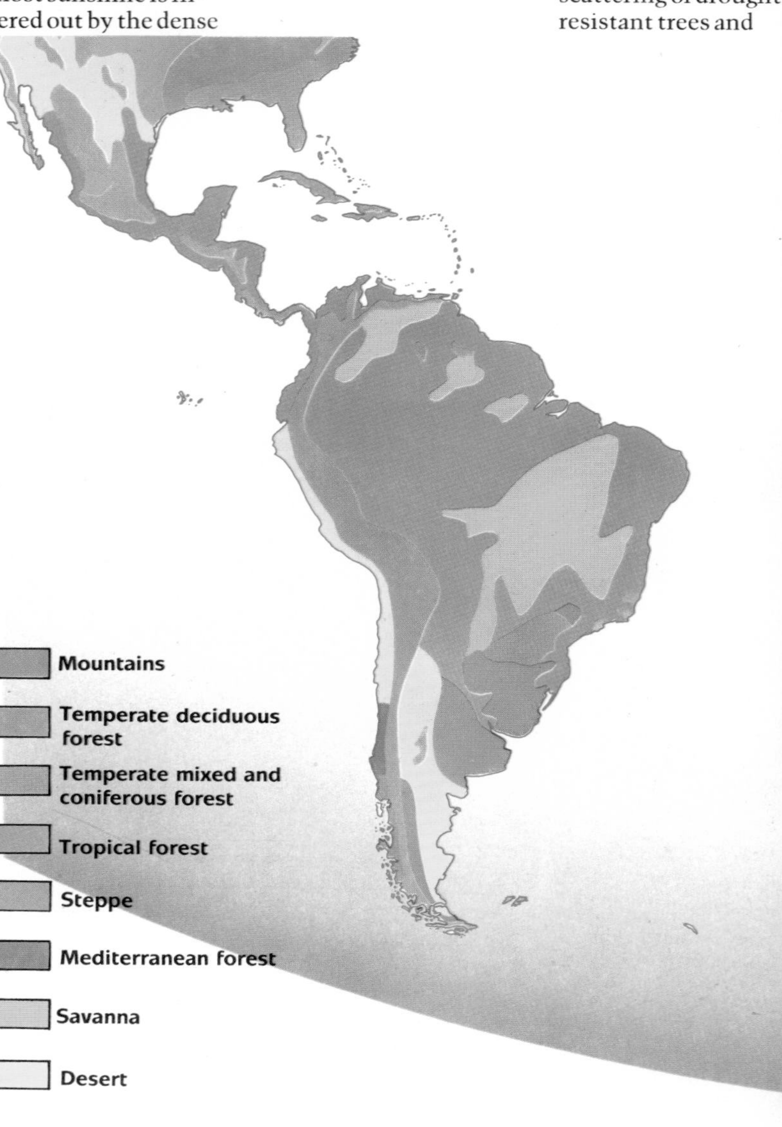

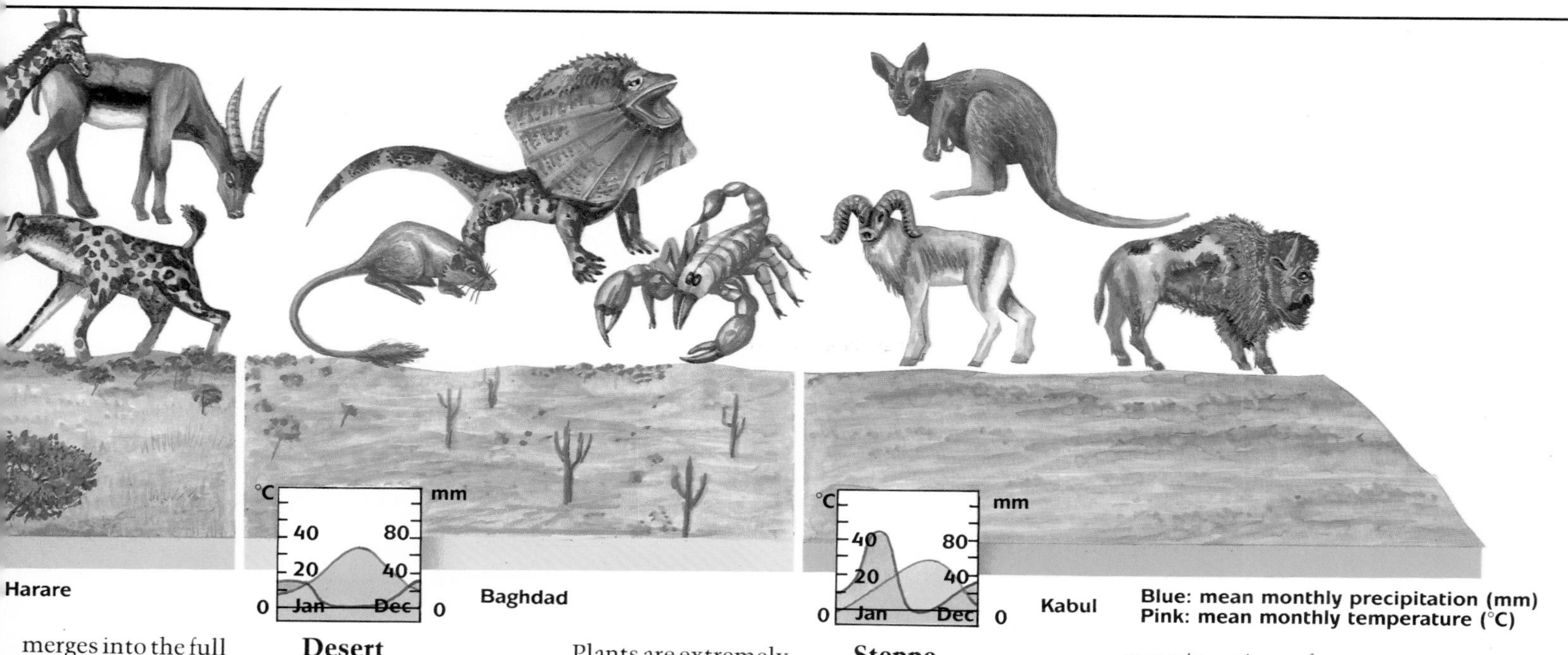


merges into the full forest. On the savanna plains roam large grazing mammals, such as gazelles, while large land carnivores, such as lions and hyenas, prey on these herds.

## Desert

In areas of less than 10 in (25 cm) of rainfall per year, even grassland vegetation cannot survive. In the desert days are hot and nights cold, as there is neither vegetation nor moisture to retain the heat. Plants are extremely resistant to drought. Seeds are watertight, while many cacti species store water in thick stems. Animals are generally small, as they must burrow to escape the heat.

## Steppe

Desert and grassland biomes can develop in temperate regions. The temperate grassland is known as prairie in North America, pampas in South America and steppe in Asia. Temperate deserts may occur in regions of extremely low rainfall. Such areas (especially the Gobi Desert of Central Asia) suffer bitterly cold winters, with temperatures down to 100 °F (40° C) below zero.

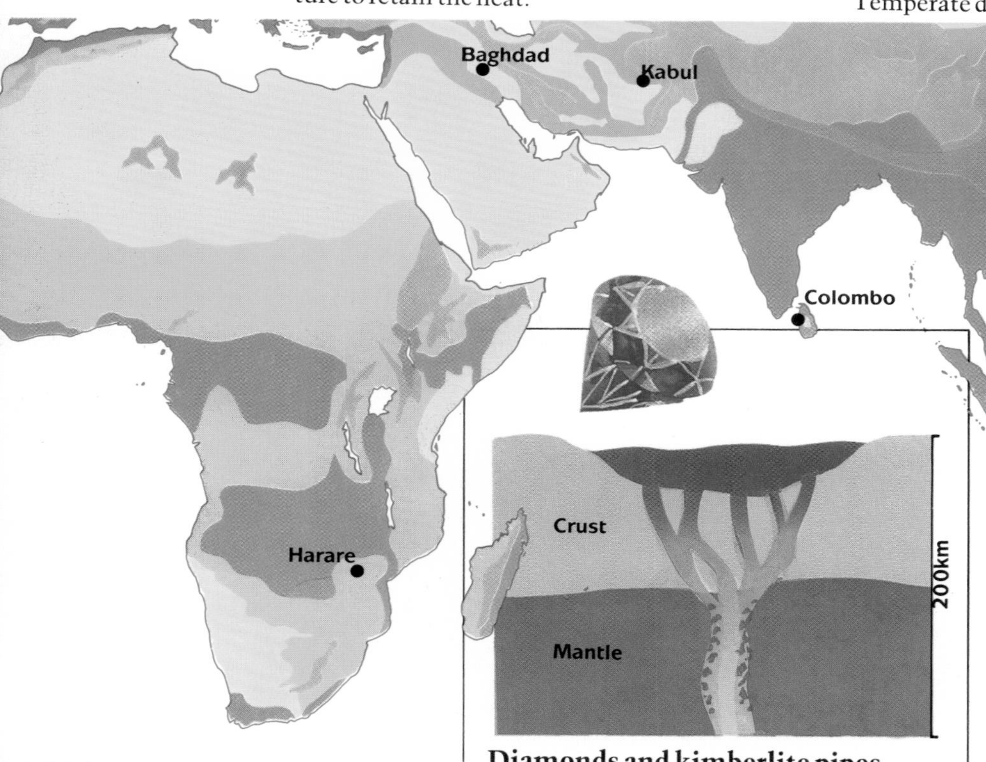


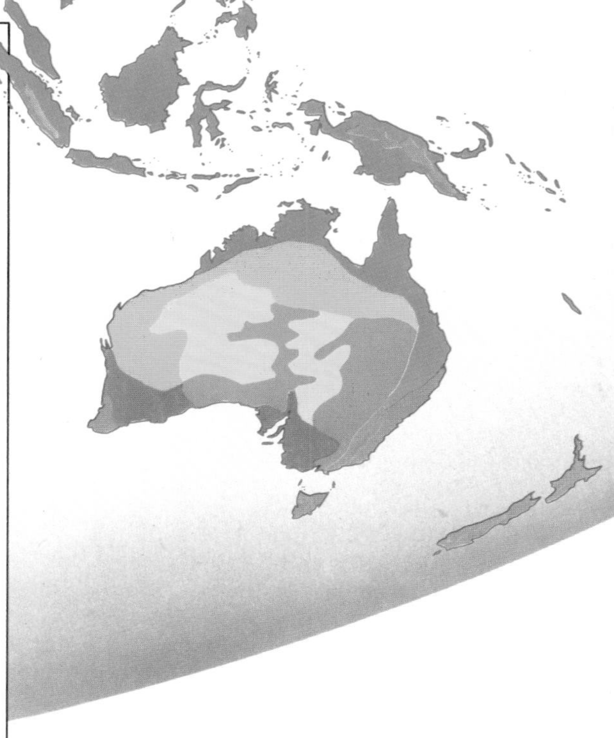

## Tablelands

Some areas of the old continental shield, in Africa and South America, exist as plateaus up to 1 mile (1.6 km) high. This phenomenon has little to do with plate tectonics: The rise of less dense mantle material has simply elevated the overlying continents. These strange plateaus include the raised tablelands of northern Brazil once thought to preserve long-extinct animal species.

## Diamonds and kimberlite pipes

Liquid accumulating in the undisturbed mantle beneath continental shields may rise towards the surface, carrying with it rocks from depths of up to 120 miles (200 km). At such depths carbon atoms are compressed into diamond, the hardest known natural mineral. Diamonds are found within the pipes leading up from the mantle infilled with a rock known as kimberlite.

EVOLVING ENVIRONMENTS

# The Great Rift

Running south from eastern Turkey a great fractured valley, or rift, in the crust of the continent can be followed for more than 3,000 miles (5,000 km). It cuts through the lands of the Bible to where the Red Sea has torn Arabia from Africa, and then branches through eastern Africa until it finally disappears in the southeast edge of that continent. The lakes, rivers, and mountains along the Great Rift have provided many habitats where life has evolved.

The northernmost 600 miles (960 km) of the rift comprises a major fault zone subject to horizontal movements. At places where the line of the fault has shifted, deep depressions have formed, such as the Sea of Galilee and the Dead Sea. Further to the south, upwelling of the mantle causes the zone of weakness along which the rift occurs. The rise of mantle material that creates the plateaus to either side of the rift is indicated by the volcanoes that are distributed along the whole length of the Great Rift.

**The Dead Sea**
The Dead Sea "pull-apart" is, at 1,300 ft (390 m) below sea level, the lowest point on the land surface of the continents. Rapid evaporation and the low rate of flow of the River Jordan has prevented it from filling with water.

**The cradle of man**
The earliest hominid, known as *Australopithecus*, may have evolved on the dry grasslands of the East African Rift Valley. A partly complete human fossil from southern Ethiopia, and footprints preserved in Kenya, confirm that humans were walking upright by 3.3 million years ago.

The first true human, *Homo habilis*, evolved from the more slender *Australopithecus* species some 2.5 million years ago. Around 1 million years ago both species died out, giving way to the more advanced *Homo erectus*. He grew to heights more typical of modern humans and made stone tools. He spread from Africa into Asia and Europe and later to Australia and North America.

## The Red Sea

To the south the Dead Sea rift passes into the Red Sea, a new ocean in the making. Down the center of the Red Sea rift runs a spreading ridge that has been separating Arabia from Africa for the past 10 million years.

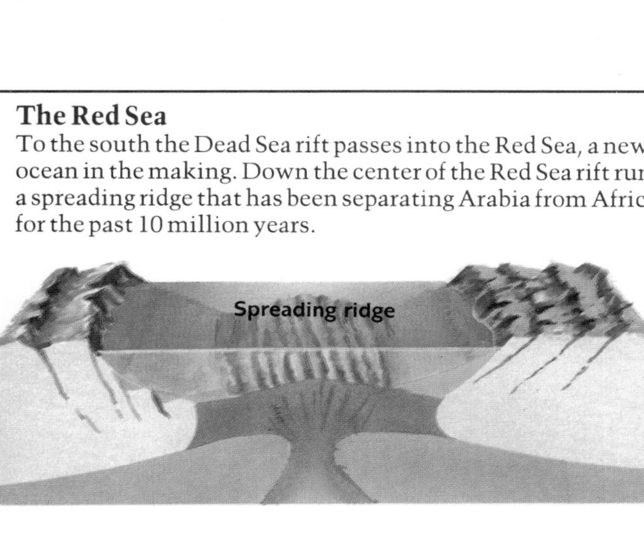


## East Africa

Farther to the south the East African Rift Valley resembles the Red Sea before the ocean crust welled up in the middle. From the Danakil depression, the rift passes through the uplifted lava plateaus of Ethiopia, then forks in Kenya.

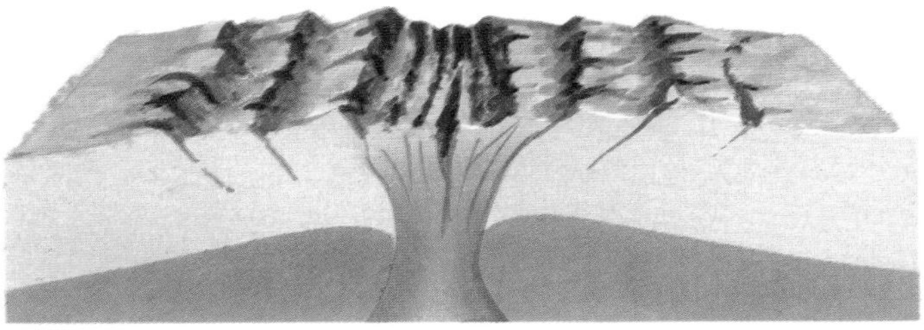

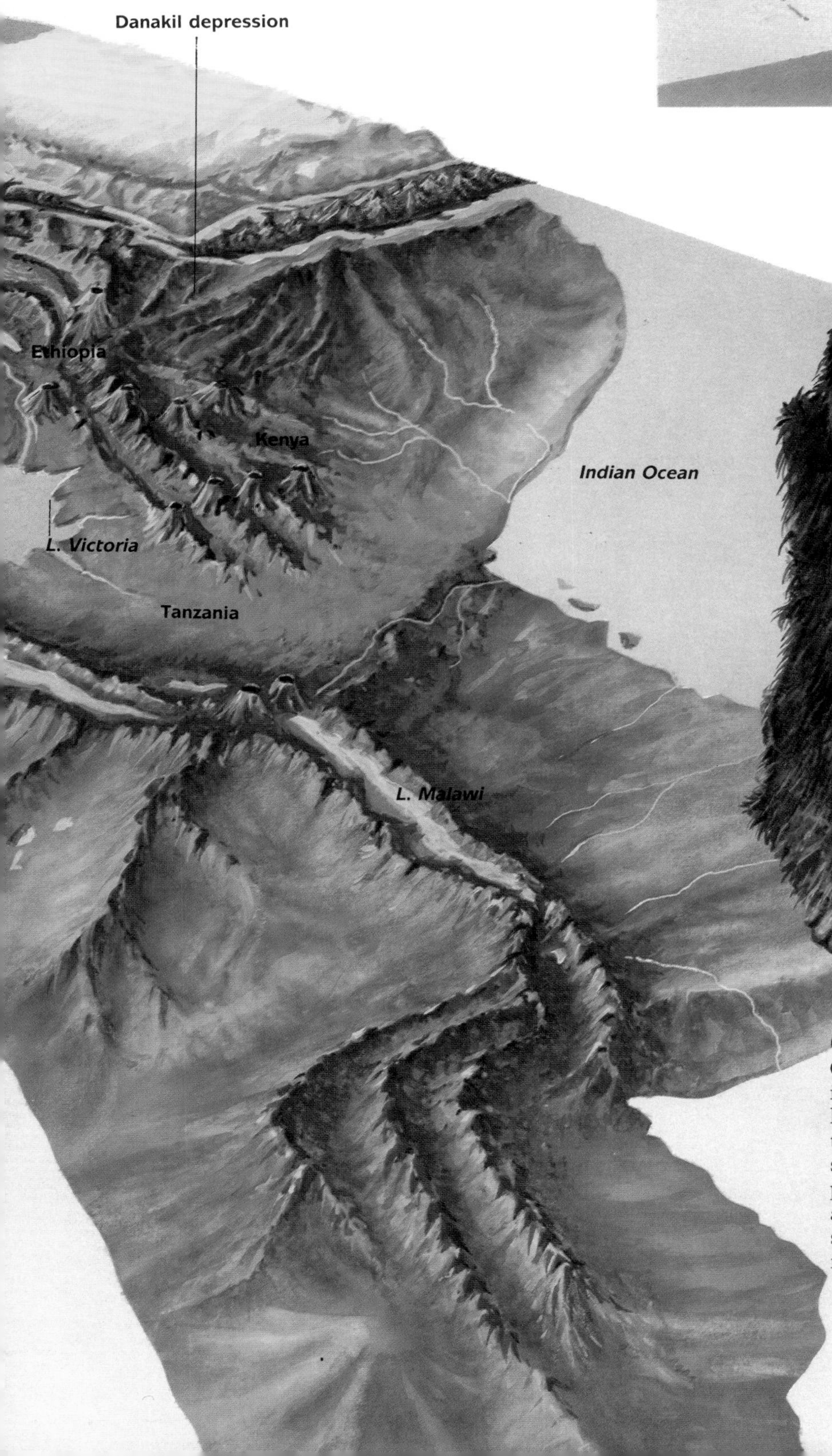

## Gorillas

On some of the highest mountains along the rift in equatorial Rwanda and Zaïre, in the thick lush rain forests, lives the gorilla. It is likely that humans, chimpanzees, and the gorilla share a common ancestor; the three species probably separated only around 5 million years ago. Gorillas share a similar life expectancy and stages of development to humans, but adult males can reach weights up to 600 lb (270 kg).

# Faults and sediments

When the crust of the continents becomes stretched, it opens up like a muscle. At the base of the crust the rocks are hot and sticky, whereas in the top 9 to 12 miles (14 to 19 km), the rocks are hard and brittle. When they are stressed, they break. When the crust is torn apart, it opens up along a series of fractures, dipping at a steep angle. These are known as extensional faults.

Extensional faults play an important role in the deformation of the continents. Once they have been formed they become repeatedly reused, from one episode of fault movement to the next. Even if the continent becomes stretched or pushed, after many hundreds of millions of year, faults of the right orientation may come back to life again.

Movement on extensional faults is one of the principal causes of land submergence. As the land passes underwater, so it accumulates sediments, principally the debris from the erosion of nearby landmasses but also the remains of plants and animals. As these organic materials become more deeply buried, so they are transformed into some of the most widely sought and most important geological materials. These are coal, oil, and natural gas, compounds composed principally of hydrogen and carbon, and known collectively as hydrocarbons.

An active normal fault (as in the Basin and Range, western US), lying at the boundary between a rising mountain front and an accumulating pile of sediments.

Basin and Rang
Sediments
Crystalline rock

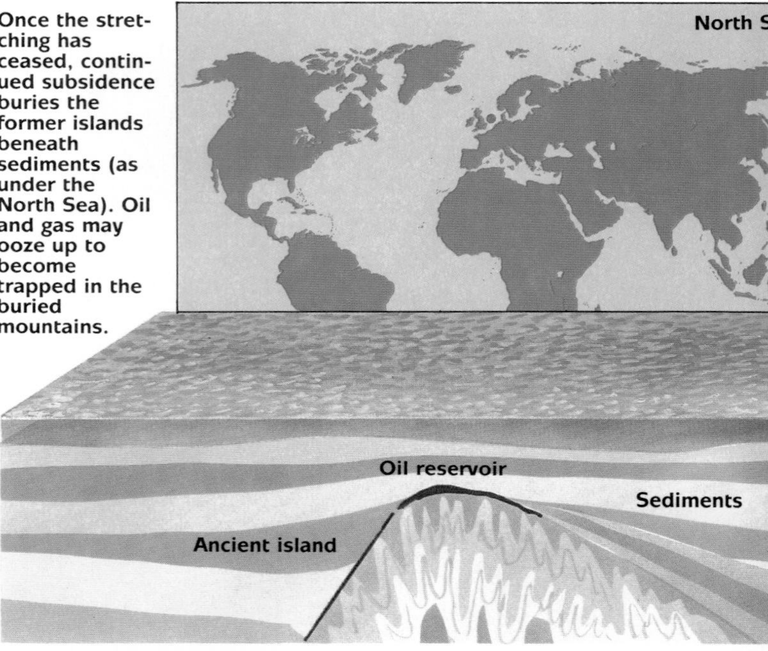


Once the stretching has ceased, continued subsidence buries the former islands beneath sediments (as under the North Sea). Oil and gas may ooze up to become trapped in the buried mountains.

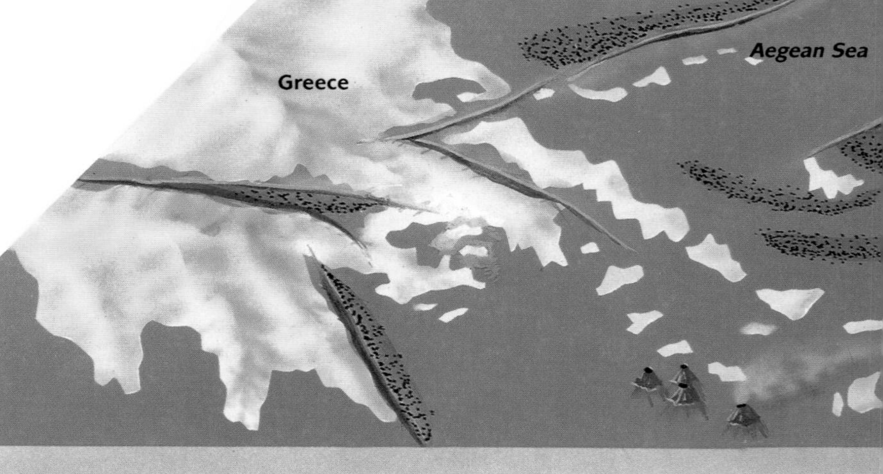


The Aegean Sea between Greece and Turkey is being stretched north-south at around two inches (50mm) per year, the crust opening up along active normal faults that bound mountains and islands.

Volcano

Extensional fault

Sedimentary basins (land lowered between faults)

Hot spring

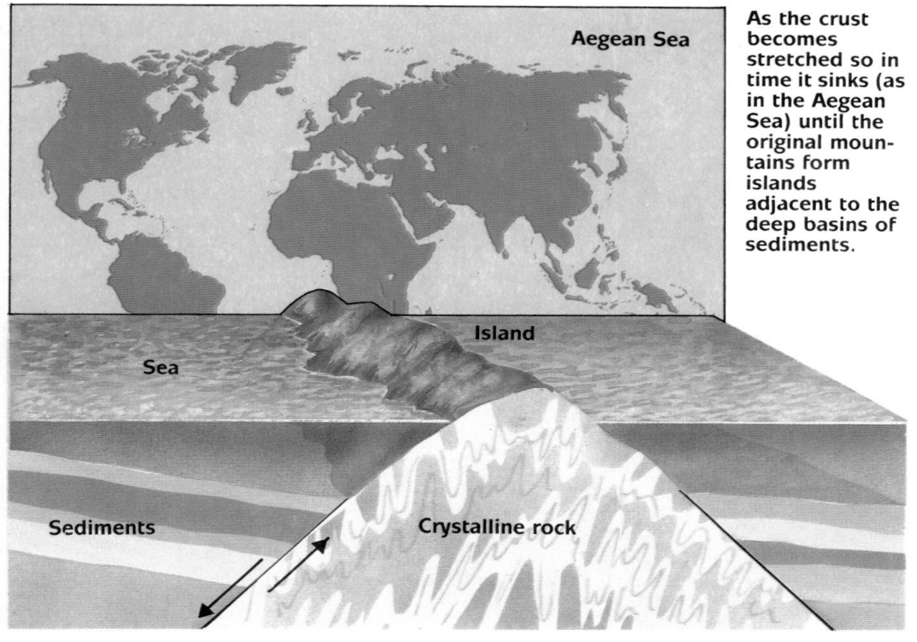


As the crust becomes stretched so in time it sinks (as in the Aegean Sea) until the original mountains form islands adjacent to the deep basins of sediments.

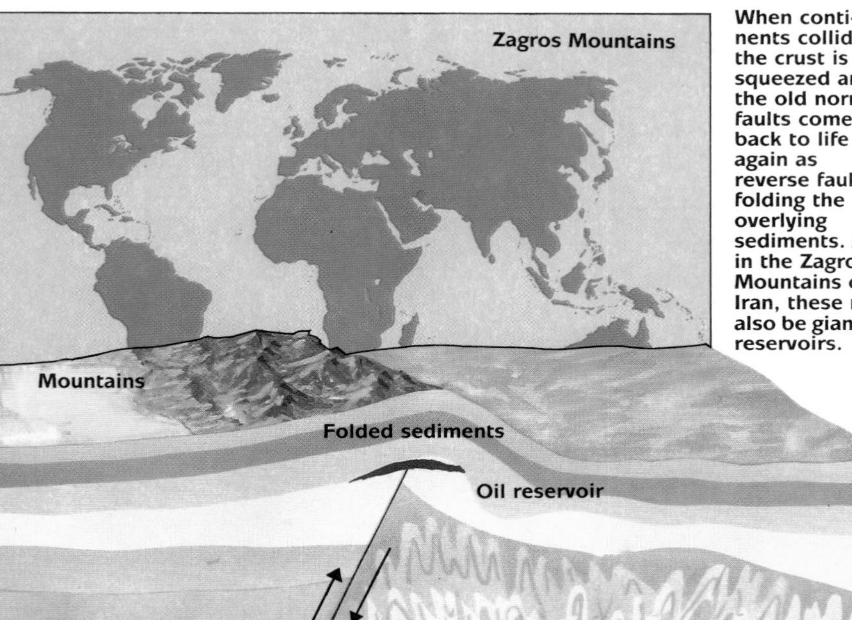


When continents collide the crust is squeezed and the old normal faults come back to life again as reverse faults, folding the overlying sediments. As in the Zagros Mountains of Iran, these may also be giant oil reservoirs.

Turkey

## Basin and Range

Two regions where the crust is being stretched today are in the Basin and Range area of western United States, and the Aegean Sea, in the Mediterranean. The name "Basin and Range" reflects the landscape: the basins are the downfaulted areas of lowland in which thick masses of sediment are accumulating; the ranges are the uplifted rocks on the other side of the fault. The lowest point in the United States, at Death Valley, is in a basin hard up against one of these major faults.

While the Basin and Range crust has been stretched, the whole region has also been uplifted by the arrival of a great mass of hot mantle material beneath the surface. The Colorado River has cut the Grand Canyon more than 5,000 ft (1,500 m) deep as the region through which the river flowed has been uplifted.

In the Aegean, the continental crust has been stretched so far it is only half its original thickness. In consequence the topography has sunk, turning the ranges into islands.

The next stage of this process can be seen in the North Sea between Britain and Norway. One hundred and fifty million years ago, it resembled the Aegean, with steep-sided islands rising out of a shallow sea. The region has since cooled and sunk, accumulating a blanket of sediment that smothers the islands.

## Oil

Most oil originates in the accumulation of the tiny ocean plants, phytoplankton. In the absence of oxygen, their bodies of these organisms are preserved from decay: they sink and accumulate on the floor of the sea. In parts of the crust where there are upside-down hollows capped with impermeable materials, "reservoirs" of oil and gas can accumulate.

These can occur in the faulted "ranges" trapped beneath a blanket of sediments, as in the North Sea. Where the crust has become pushed, and the extensional faults have begun to reverse their direction of movement, the overlying sediments become folded. In the Zagros Mountains of southwest Iran, the collision of Arabia with Asia has created extensive oil fields.

EVOLVING ENVIRONMENTS

# Continental shelves

The sunny surface waters of the seas are an ideal environment for life. Yet to live in this environment organisms have to be able to cling tight to a rigid base, in order to withstand the battering of the waves. Not content simply to rely on the rocky outcrops of a natural shoreline, living things have developed the ability to construct platforms, known as reefs, to which they can fix themselves. This ability also allows the organisms to modify their growth according to rapid changes in sea level, as accompanied the last Ice Age.

The most prolific underwater builders today are tiny coelenterates, or corals. Their achievements are awesome: the Great Barrier Reef in Australia, for example, stretches for 1,250 miles (2,000 km). Waves breaking over the coral reefs help to oxygenate the water and allow the coral to flourish. The intricate shape of reefs, filled with caves and crannies, encourages a great concentration of life, making this one of the richest environments on Earth – the marine equivalent of the rain forest.

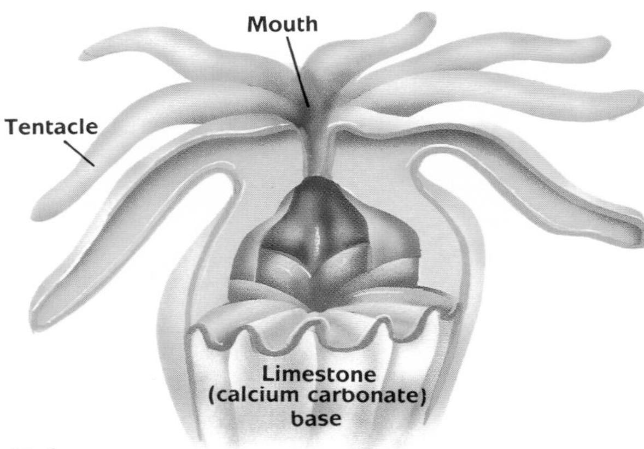


**Polyps**
Coral is a type of primitive animal known as a polyp, which secretes a rigid skeleton of calcium carbonate tubes from its base. The tubes become entangled, creating a strong limestone rock with a living surface. There are spheroidal corals, elkhorn corals, even brain corals, each with a different design and different environmental requirements.

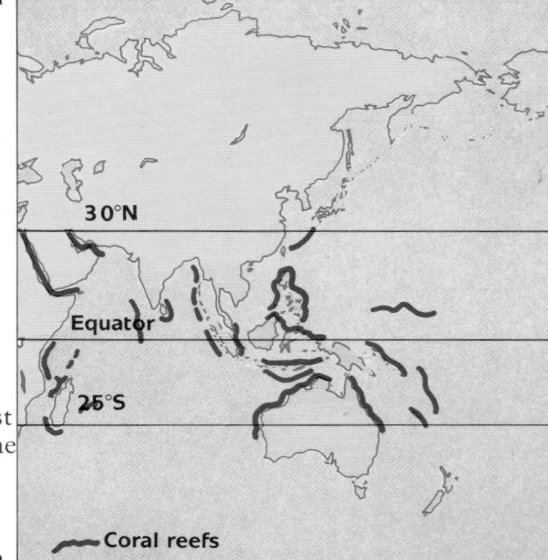


**Coral reefs**
Coral reefs occur in a narrow tropical band, from 30°N to 25°S. Although corals are animals, they also contain algae, which need sunlight. Coral, therefore, grows fastest just a few feet below the surface. Corals are extremely sensitive, thriving best in water temperatures between 77°–84°F (25°–29°C).

## Volcanic island atolls

After a volcano has become extinct, it inevitably sinks, both because the volcano itself contracts as it cools but also because the ocean floor too is cooling and sinking. If the volcano is at the correct latitude, a coral reef will develop in the surrounding shallow water becoming a ring, known as an atoll. As the volcano sinks the reef continues to receive enough sunlight to continue building for millions of years on the sinking volcano's foundations. Some coral reefs are now towers, more than 3,300 ft (1,000m) high, built on sunken volcanoes. If plate movements, however, cause the atoll to drift out of the tropics, the reef dies and the island drowns.

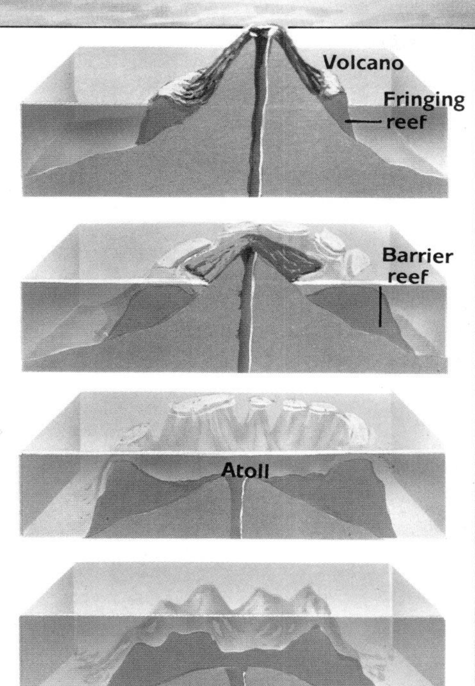


Continental shelf seas

Mississippi delta

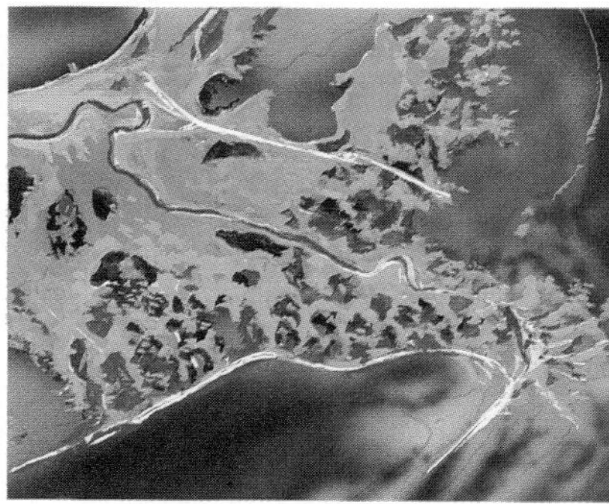

Barrier beaches

## The continental shelf

Between the land and the ocean, in almost every part of the world, lie shelf seas, generally less than 500 ft (150 m) deep. Sometimes no more than narrow ledges 6 miles (7.2 km) wide, as fringing lips of the ocean, they may reach widths of 60 miles (72 km) or more. In the geological past such widespread shallow seas were not nearly as abundant as they are today. They are a unique feature of the modern world.

The enormous expanse of the shelf seas owes its existence to the last Ice Age. So much water had become frozen into the expanded ice caps, that the global sea level was lowered by more than 100 ft (30 m). The sea scoured the newly exposed land around the continents. When the ice caps melted and sea levels rose again, these lowlands became the continental shelves.

The rapid rise in sea level created some unique coastal landscapes, reflecting the youth of the coastline. These include drowned valleys of rivers that formerly flowed into the lower Ice Age sea and barrier beaches. The great sediment-laden rivers of the world, such as the Mississippi, have had only 6,000 years in which to build new deltas out into the sea. These are pale shadows of the deltas of the past, when the sea level remained constant for millions of years. Then, deltas advanced across the whole continental margin to the edge of the deep ocean.

# Volcanoes

When the white-hot interior of the Earth breaks through to the surface it can have the most devastating impact on the surrounding region. The fiery molten rock (known as magma) can be hotter than 1,800°F (1,000°C).

The magma arrives at the surface either through a broad pipe or along a narrower, elongated fissure. Solidified magma builds up around the vent to create the characteristic form of the volcano – high and steep where the magma is cool and sluggish, but a broad, rounded dome where the magma is hot and flows almost like water.

As the magma rises beneath a volcano, the pressure builds up until suddenly the overlying rocks break. The dissolved gas foams out of the magma, expanding the column and sometimes creating an enormous explosion. Damage comes from a whole arsenal of deadly agencies: hot toxic gas, high-speed shock waves, hurricane force winds, hailstorms of rocks, volcanic ash blizzards and colossal landslides of volcanic ash mixed with water.

**Volcanoes are located along many plate boundaries. Whereas those of the mid-ocean spreading ridges are almost always underwater, the largest and most dangerous volcanoes lie above subduction zones, in particular around the Pacific.**

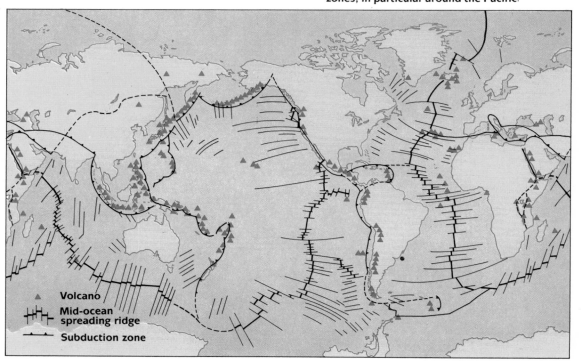

Volcano
Mid-ocean spreading ridge
Subduction zone

### Fertile volcanic soils

While everything that lies in the path of the hot magma, ash, and gas is killed, the fresh rock from the mantle contains crystals that decompose rapidly on contact with water, producing a soil full of the mineral nutrients vital for life. At a distance from a volcano, where only a few inches of ash are deposited, grass and trees quickly recover. Farmers in Italy and the West Indies used to store volcanic ash to fertilize their fields. Because volcanic soil is so fertile many people live on the flanks of active volcanoes, even though there is always the possibility of renewed eruption.

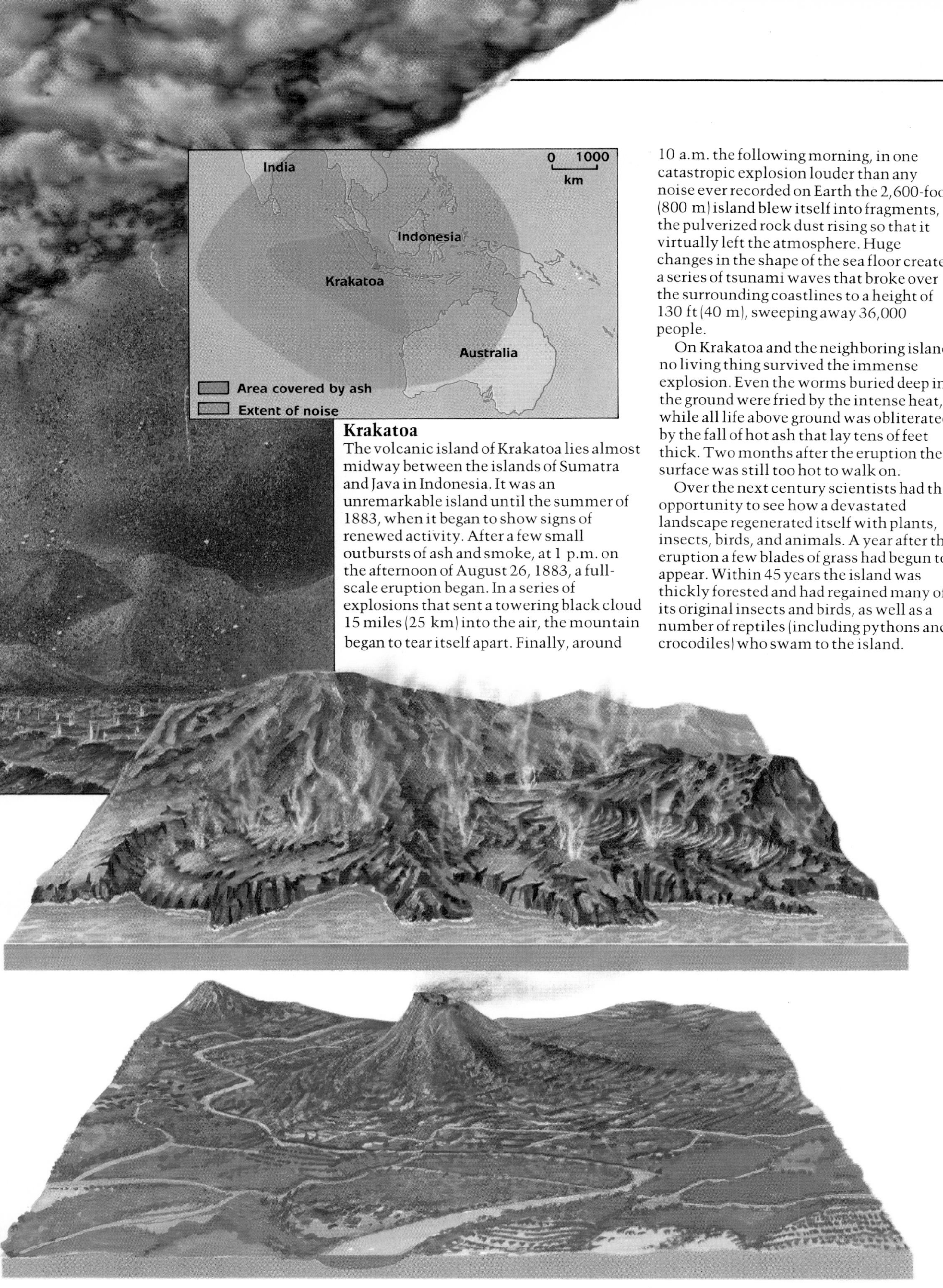


## Krakatoa

The volcanic island of Krakatoa lies almost midway between the islands of Sumatra and Java in Indonesia. It was an unremarkable island until the summer of 1883, when it began to show signs of renewed activity. After a few small outbursts of ash and smoke, at 1 p.m. on the afternoon of August 26, 1883, a full-scale eruption began. In a series of explosions that sent a towering black cloud 15 miles (25 km) into the air, the mountain began to tear itself apart. Finally, around 10 a.m. the following morning, in one catastropic explosion louder than any noise ever recorded on Earth the 2,600-foot (800 m) island blew itself into fragments, the pulverized rock dust rising so that it virtually left the atmosphere. Huge changes in the shape of the sea floor created a series of tsunami waves that broke over the surrounding coastlines to a height of 130 ft (40 m), sweeping away 36,000 people.

On Krakatoa and the neighboring islands no living thing survived the immense explosion. Even the worms buried deep in the ground were fried by the intense heat, while all life above ground was obliterated by the fall of hot ash that lay tens of feet thick. Two months after the eruption the surface was still too hot to walk on.

Over the next century scientists had the opportunity to see how a devastated landscape regenerated itself with plants, insects, birds, and animals. A year after the eruption a few blades of grass had begun to appear. Within 45 years the island was thickly forested and had regained many of its original insects and birds, as well as a number of reptiles (including pythons and crocodiles) who swam to the island.

**Volcanic landscapes are extremely fertile because the minerals in ash and lava rapidly decay to produce thick, rich soils. The towns and villages that gather around the volcano must live with the constant threat of a catastrophic eruption that turns this landscape back into a barren wilderness.**

NATURE IN CHANGE

# Earth movements

Changes in the level of the Earth's crust can occur gradually and imperceptibly as a result of slow underlying movements of the mantle. They may also occur suddenly in association with earthquakes.

Earthquakes are simply the vibrations felt on the surface when buried rocks break. Just as a breaking log sounds louder and deeper in pitch than a snapping twig, so the bigger the area of rock that breaks, the louder and deeper the vibration. The vibration of most earthquakes is so deep-pitched, in fact, that it cannot be heard – only felt. Such vibrations can shake buildings to destruction. The places where the rocks break are known as geological faults.

Zones of concentrated earthquake activity, where enormous rock masses collide or slide past one another, define the boundaries of the Earth's plates. However, there is also slight internal movement within the plates, which are not always completely rigid. The largest recent earthquake in the continental United States occurred far from any plate boundary, on the border between Tennessee and Missouri, December 1811.

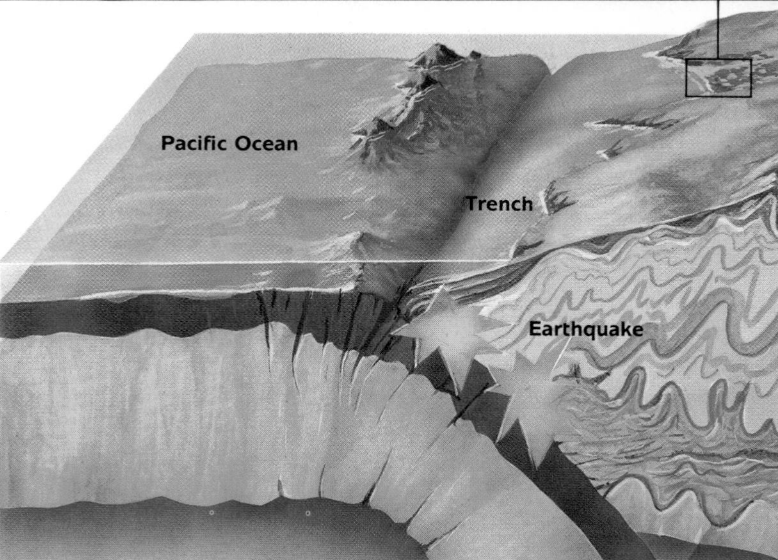


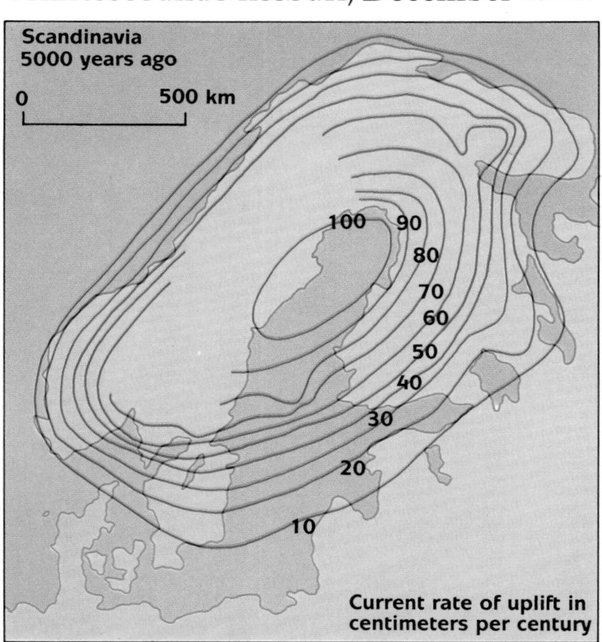


### Changes in land level

Northern Scandinavia is a remarkable area, one subject to changes in land level without any serious accompanying earthquakes. Only 14,000 years ago an ice cap 10,000 feet (3,000 m) thick covered the area weighing down the crust, which sank into the softer underlying mantle. The ice cap melted but the crust did not recover immediately. It continues to rise as fast as four-tenths of an inch each year.

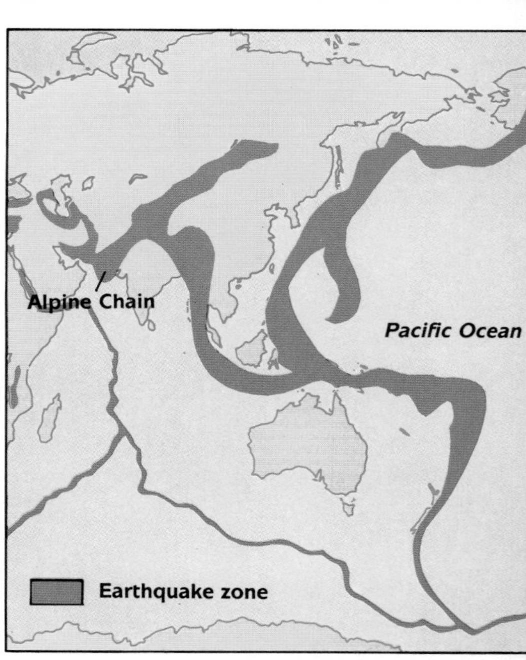

## The great Chilean earthquake

On May 22, 1960, a piece of rock broke somewhere beneath the coast of central Chile. The crack continued to grow wider and longer. About five minutes later it ceased, having traveled about 600 miles (960 km). It was the largest earthquake of the past two centuries. The crack lay on the boundary between the Pacific (Nazca) and South American plates, where the oceanic crust begins to curve down into the mantle, beneath the Andean mountains.

As the crack expanded, the rock to either side moved by a total of around 65 ft (20 m), as the western edge of South America advanced westward over the ocean crust. This set off enormous vibrations that shook buildings until they disintegrated.

The movement of the sea floor also created a series of giant waves, known as tsunamis, that rose to heights of up to 60 ft (18 m), sweeping away low-lying villages and destroying coastal cliffs and beaches.

This great wave also spread out into the Pacific, moving almost as fast as a jet plane. In the middle of the night, a wall of water 50 ft (15 m) high burst over the town of Hilo on Hawaii 5,000 miles (8,000 km) away.

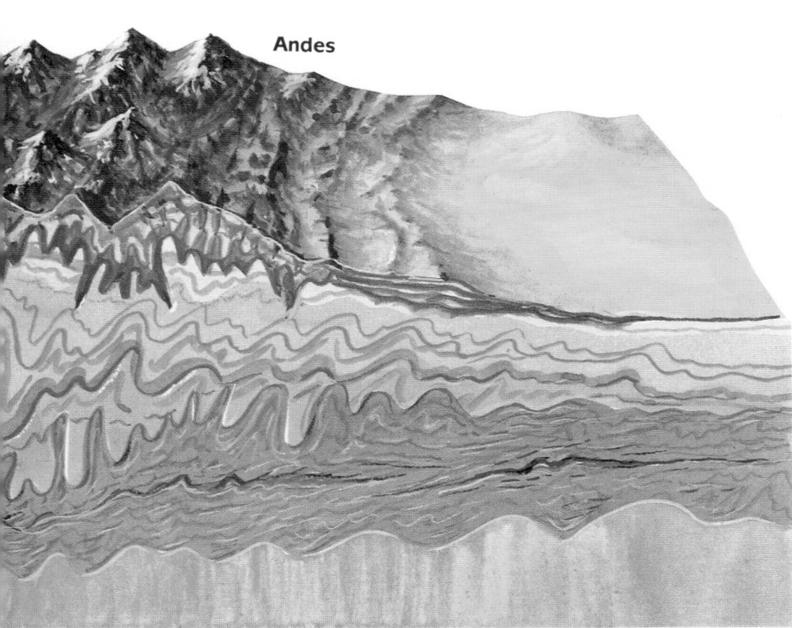


## Big fault breaks

Big fault breaks, those that cause major earthquakes, pass through to the surface. Where the rocks are in compression and one side of the fault overrides the other, a "reverse fault" is formed. Where the rocks are in extension and one side of the fault moves down relative to the other, there is a "normal fault." Where the two sides of the fault move horizontally, there is a "strike-slip fault." While reverse faults tend to lift land up, normal faults do the opposite. Strike-slip faults, such as the San Andreas Fault in California, cause river channels to become offset, and create a curious series of ponds and mounds along slight bends in the fault line.

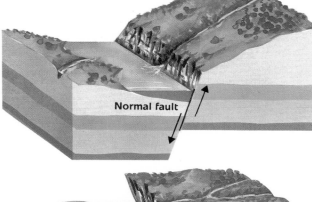

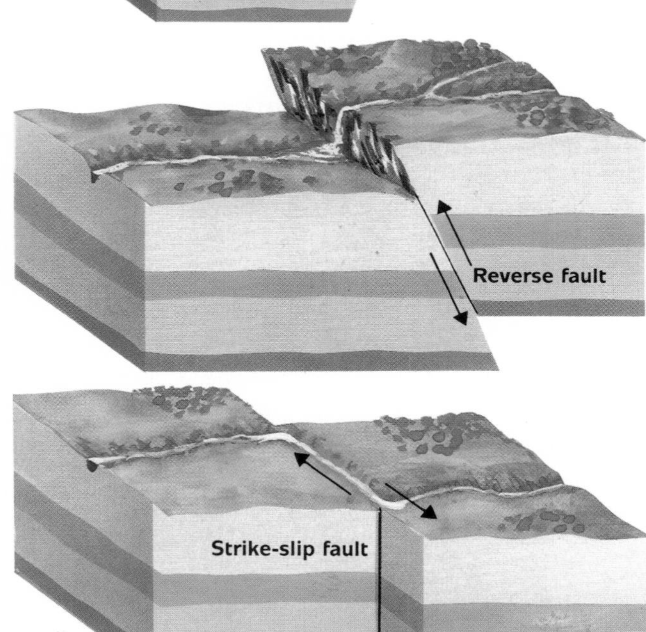


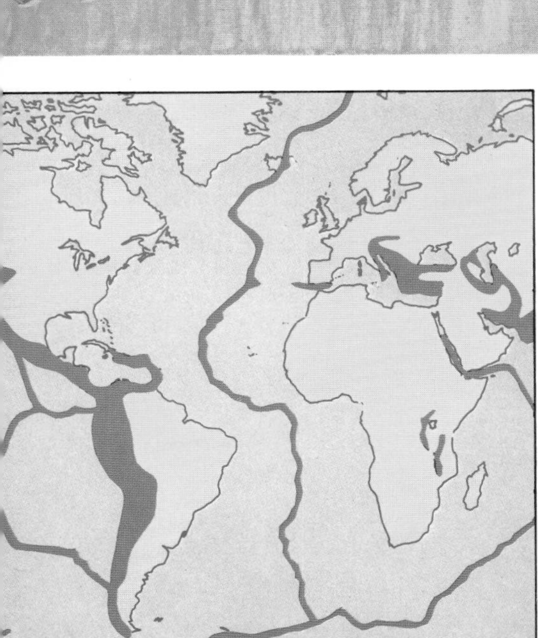

## Earthquake distribution

The distribution of earthquakes around the world provides a snapshot of where rocks are breaking most rapidly. 90 percent of earthquakes occur in the "Ring of Fire" which rings the Pacific Ocean. Many others occur in the Alpine Chain, which stretches from Europe right through to Southeast Asia.

## The size of earthquakes

Earthquakes are measured on a magnitude scale, which gives an approximate indication of the energy released. On the most familiar Richter scale of magnitude, each level is around 30 times the energy of the one below. Small tremors may be magnitude 3 or 4; the great 1960 Chile earthquake was magnitude 9.5.

NATURE IN CHANGE

# Meteorite collisions

The most massive instantaneous calamities suffered by planet Earth originate from outside, in the sudden arrival of enormous meteorite boulders. Such boulders have two sources: The large population of roving asteroids that generally follow orbits between Mars and Jupiter, and the comets with highly eccentric paths that take them across the orbits of all the inner planets. It is only a matter of time before a cometary nucleus collides with a planet at a speed of up to 40 miles per second.

The amount of heat-energy liberated on impact becomes so great that all but fragments of the meteorite are vaporized. Therefore, the largest collisions in the geological past are known only from the destruction that they wrought: first, the hole smashed in the Earth's crust; second, the more distant effects, particularly on life, from powerful shock waves, and a thick cloud of dust encircling the globe.

Meteor Crater

**Ries crater**
Meteor Crater, Arizona (left), is at 1 mile (1.6 km) across dwarfed by the Ries crater, near Munich in West Germany. Formed 15 million years ago, it measures 15 miles (24 km) in diameter. Fractured crustal rocks continue to a depth of 5 miles (8 km), overlaid by 1,300 ft (390 m) of rock smashed to such an extent that it has melted. A characteristic feature of impac craters such as the Ri is the uplifted center the crust beneath we up to fill some of the space left by the impact.

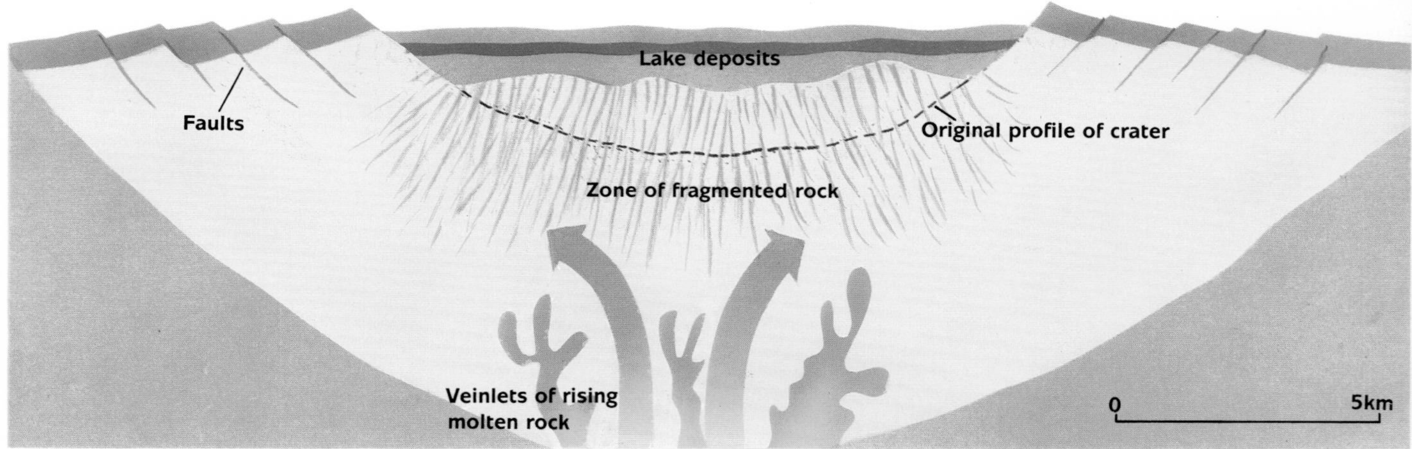

## The end of the dinosaurs

The remarkable reign of the dinosaurs came to an abrupt end 65 million years ago, when in a relatively short time all varieties of dinosaur were annihilated. However, many other species also disappeared, including winged reptiles, the large marine reptiles, the last ammonites, and an estimated 15 percent of all marine animal families. It is interesting to note the fate of the survivors: on land smaller animals survived, including the warm-blooded mammals and birds, while in the water larger animals such as crocodiles resisted extinction. Everything points to some short-lived episode of cold when only animals that hid in burrows or in water survived.

It seems most probable that a major cause of what happened at this time was the collision with one or more giant meteorites. By this theory the dust from the collision shrouded the Earth, turning the climate into a temporary frigid "winter," even in the tropics. This killed the trees, which were then consumed by fire. The only problem is that there is evidence to suggest that it did not happen in a single incident, and more puzzling, no crater of such a collision has yet been discovered. The quartz crystals suggest that the crater should have been on the continents rather than in the oceans: perhaps in a region of Antarctica or Greenland now covered with ice?

**Craters pepper the surface of the planet. The oldest continental shields contain the highest concentration of large impact craters: there are 29 known craters on the North American shield.**

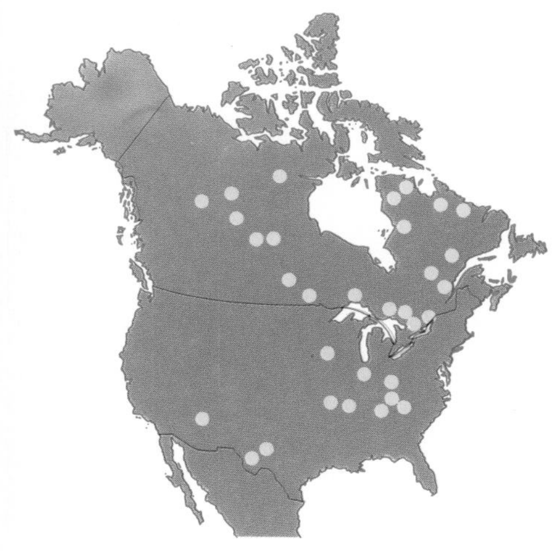

## The Tunguska fireball

The most spectacular encounter with an object from outer space so far this century occurred without warning at 7.17 a.m. local time on June 30, 1908, around the Tunguska river valley in a remote region of northern Siberia. An enormous fireball and explosion devastated the forest; trees were consumed by the heat up to 9 miles (14 km) from the center, and felled by the blast up to 30 miles (48 km) away. A farmer 36 miles (58 km) away described "a great flash of light and so much heat that my shirt almost burned off my back. A huge fireball covered an enormous part of the sky, followed by an explosion that threw me several feet from the porch." The event described is believed to have been a collision with a comet, exploding at an altitude of between 3 and 5 miles. Fortunately this was one of the least populated regions in Asia and the only casualties were some 1,000 reindeer. The implications of such an explosion over Moscow, London, or New York can only be guessed at.

# Extreme weather

While plants and animals are adapted to withstand a wide range of temperatures, rainfall, and air movements, they have little protection against infrequent, immensely destructive weather systems. The most powerful of these are the tropical storms, the hurricanes.

Like every other part of the Earth's great weather engine, tropical storms are powered by temperature differences. A huge column of warm, moist air rises, and condenses into clouds, releasing enormous quantities of heat. This heat powers the storm's giant spinning wheel of cloud, wind, and rain.

Tornadoes, twisting columns of air that reach to the ground, form as the low pressure heart of a rotating storm sinks, spinning faster as the cloud column narrows. Tornadoes generally touch ground for just a few miles, gliding over the surface at about 30 mph (48 kph), although, in the vortex itself, wind speeds can reach 270 mph (430 kph) often leaving behind a trail of destruction.

**Floods**
After heavy rainfall, rivers may overflow their banks and, if unchecked, spread out over the floodplain. Many rivers change their courses quite naturally, often as a result of such floods. Floods are a natural part of the evolution of landforms and the creation of new land.

Deforestation, as in the Himalayas, increases the speed at which rain runs off the land, raising the mud level in the river. Downstream in the Ganges delta, the river becomes choked with mud, raising the floods still higher.

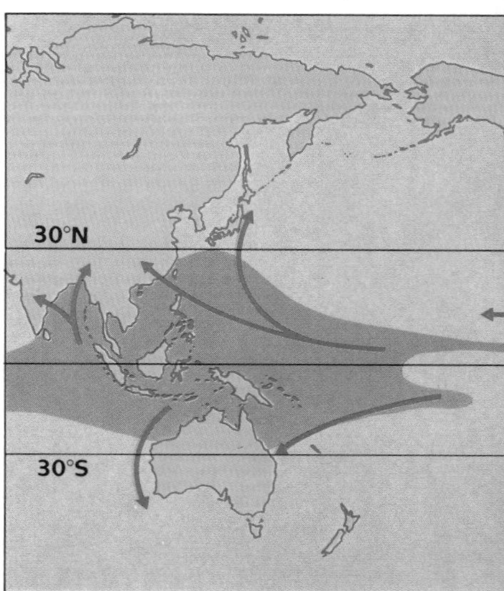


**Hurricane paths**
The seeding grounds of tropical storms are the same from one year to the next – over the warm, tropical oceans. The curved paths the storms take are also fairly

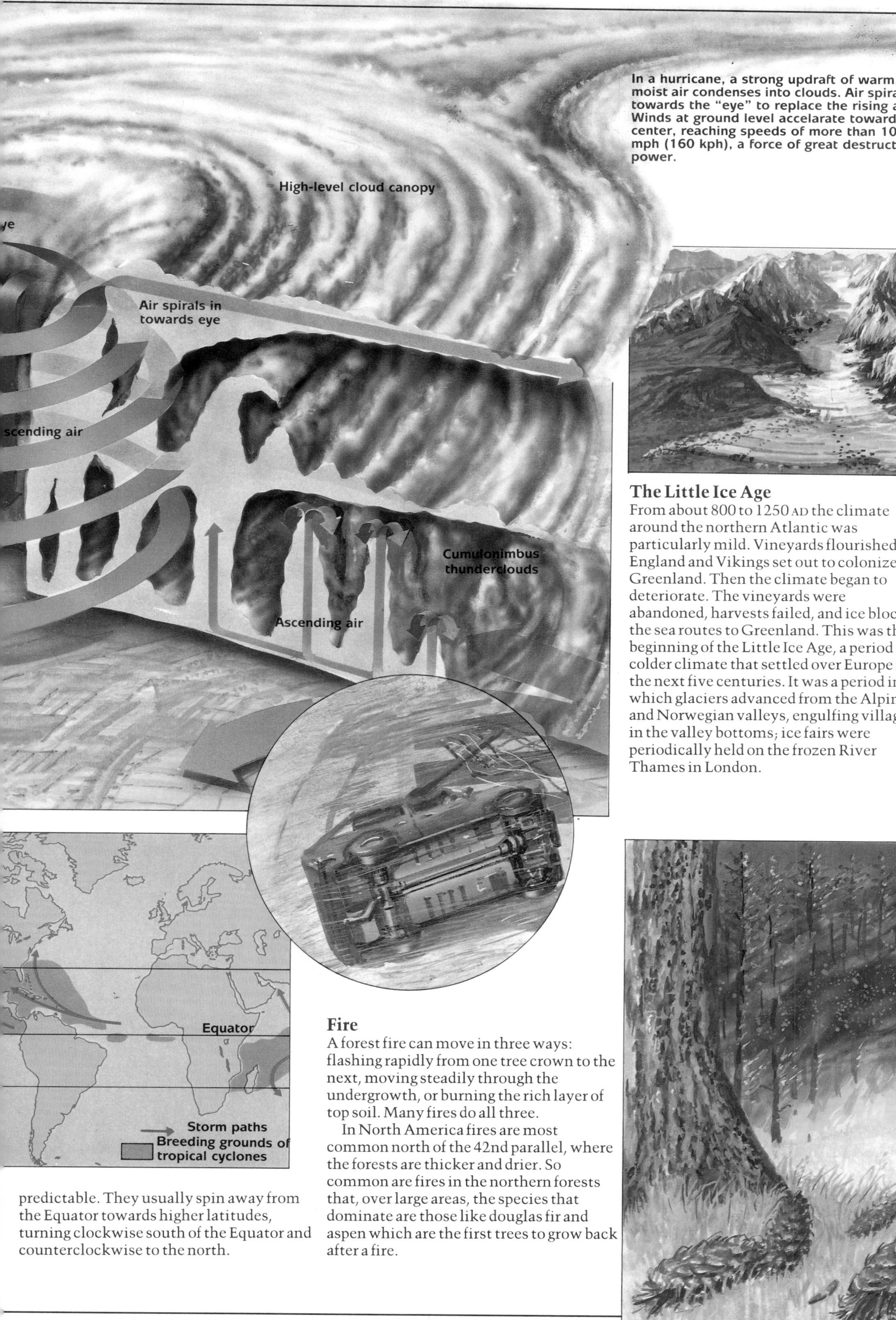


predictable. They usually spin away from the Equator towards higher latitudes, turning clockwise south of the Equator and counterclockwise to the north.

In a hurricane, a strong updraft of warm, moist air condenses into clouds. Air spirals in towards the "eye" to replace the rising air. Winds at ground level accelarate toward the center, reaching speeds of more than 100 mph (160 kph), a force of great destructive power.
High-level cloud canopy
Air spirals in towards eye
Cumulonimbus thunderclouds
Ascending air
Equator
Storm paths
Breeding grounds of tropical cyclones

## Fire

A forest fire can move in three ways: flashing rapidly from one tree crown to the next, moving steadily through the undergrowth, or burning the rich layer of top soil. Many fires do all three.

In North America fires are most common north of the 42nd parallel, where the forests are thicker and drier. So common are fires in the northern forests that, over large areas, the species that dominate are those like douglas fir and aspen which are the first trees to grow back after a fire.

## The Little Ice Age

From about 800 to 1250 AD the climate around the northern Atlantic was particularly mild. Vineyards flourished in England and Vikings set out to colonize Greenland. Then the climate began to deteriorate. The vineyards were abandoned, harvests failed, and ice blocked the sea routes to Greenland. This was the beginning of the Little Ice Age, a period of colder climate that settled over Europe for the next five centuries. It was a period in which glaciers advanced from the Alpine and Norwegian valleys, engulfing villages in the valley bottoms; ice fairs were periodically held on the frozen River Thames in London.

NATURE IN CHANGE

# Man and the environment

Ten thousand years ago, as the Ice Age ended, one species, no longer content to adapt to the environment, began to adapt the environment to suit itself. The species was man. The principle of modifying the environment became so successful that, in a mere instant of geological time, it was to transform the face of much of the planet.

Like all animals the most important determinant of early man, of where he lived and how he organized his tribes, was the search for food. From small beginnings, probably in the Middle and Far East, man learned that certain wild food plants could be grown under supervision in small, carefully tended plots of land. This marked a fundamental shift from a nomadic existence to the beginning of settlement.

At about the same time, animals too began to be domesticated, trained, and bred to share the company of man, and in exchange for food, to do his bidding. Wolves were tamed first and became the forerunners of dogs. By similar programs of breeding, wild sheep, goats, pigs, water buffalo, and birds like turkeys, ducks, and chickens were domesticated.

Early Neolithic agriculture in Europe around 6,000 BC. At that time, man began to clear forests to make space for field crops and pastureland, and so feed his expanding population.

Kangaroo
Harvest mouse
Rabbit
Partridge

Nomadic herding
Livestock farming
Shifting cultivation

## Changing the environment

The chief victims of man's success in agriculture have been the Earth's original biomes. The first farms were simply clearings in the middle of the forest, or fields in the floodplain of a river valley. Since then, however, agriculture has proved unstoppable, driven by the ever-rising population that it has permitted. No longer controlled by the limits imposed by the natural environment, man has been able to manipulate the land's resources, seemingly without limit. Yet man has now reached a crisis point, and is in great danger of imitating other island animal species, which have over-exploited their food resources, provoking mass starvation and population collapse.

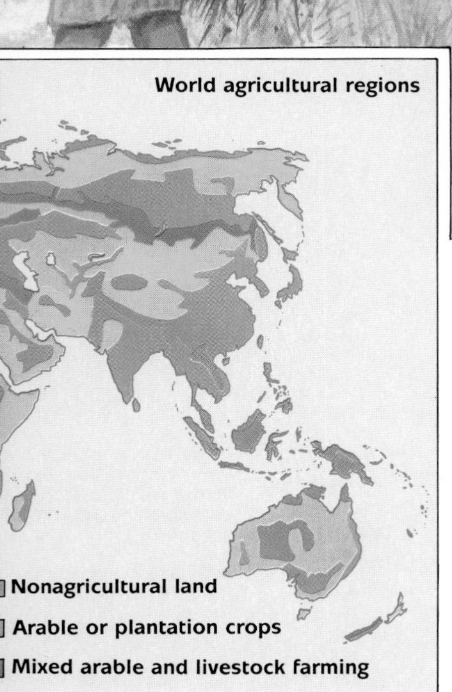


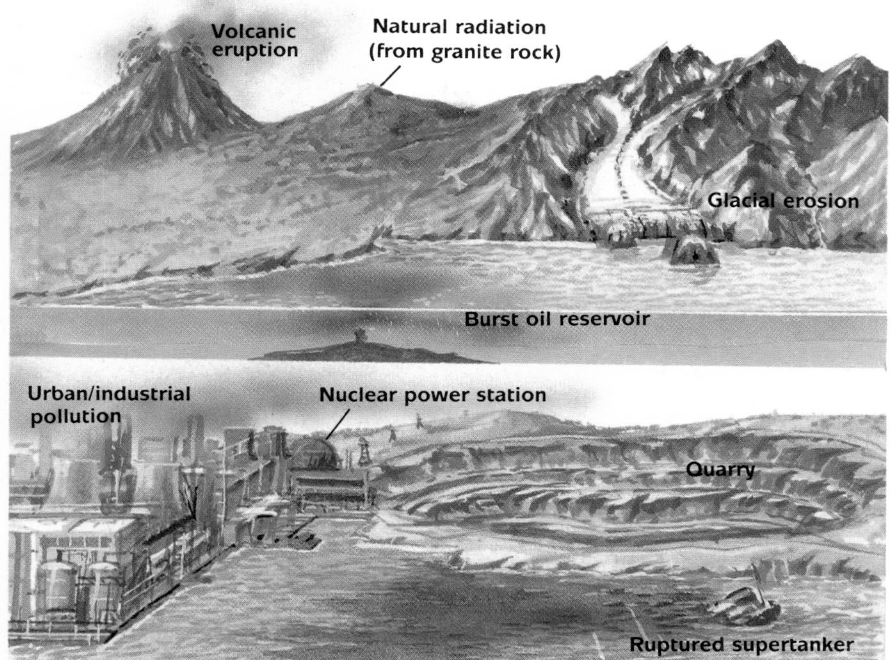


## Technology

While agriculture has proved the major force in modifying the environment, the ability to grow more food has allowed human populations to expand and this in turn has led to many other environmental changes. The larger population has to be housed in expanding towns and cities. Building materials have to be collected from quarries or forests. Above all else the demands for transport, for heating and cooling, and power for machines have all required ready supplies of energy. At least 80 percent of this energy has come from fossil plants, in the form of oil, gas, and coal. Nuclear power now supplies around 18 percent of the world's electricity.

On a geological scale, man's impact on the planet looks slighter than an individual's view of environmental change. All the world's quarries and mines are as nothing compared with the destruction wrought by the last Ice Age. In the past, natural oil reservoirs have burst, releasing a thousand times as much oil as a ruptured supertanker. Even radioactivity is "natural." When the Earth was young, surface radiation levels were ten times higher than they are today. Not even the city of Los Angeles is such an effective atmospheric polluter as a major volcanic eruption. Yet man's impact is dangerous, not because so many of the things he is doing are "unnatural," but because they are all happening together.

## Man-made biomes

Agriculture has extended across all the world's biomes, from the tundra to the tropics. As a result, man has destroyed environments, plants, and animals that had been evolving for millions of years.

The new environments that man has created, the fields and towns, have created new biomes, which have allowed certain other species to increase in number. These include rabbits and kangaroos, that benefit from arable land, crows and magpies, that now use tall buildings for nesting where once they used trees, and pigeons that roost on the artificial cliffs of the city center. As man has modified the climate of his houses, and imported all kinds of food stuffs into his domestic environment, so some tropical animals, such as the house mouse, the black rat, and the cockroach have come to live with him.

# Is the Earth alive?

Does life on Earth regulate the environment for its own benefit? Could the climate on Earth over the last 4 billion years have been prevented from becoming too hot or too cold by the active intervention of living organisms? This idea of a self-regulating planet, named after the Greek Earth goddess Gaia, was first proposed in 1972 by James Lovelock, an English chemist and inventor, working with NASA in the search for life on other planets.

Coexistence for the mutual benefit of two or more species can be seen at every level of life on Earth. Every living cell contains microorganisms (in animals the energy-transforming mitochondria; in plants the mitochondria and chloroplasts) that were formerly free-living bacteria. Yet it is a big step to suggest that the whole planet exists in some kind of active, living equilibrium. Many scientists regard the whole idea as too mystical, yet a number of components of self-regulation have been uncovered. Already Lovelock's work has provided a powerful stimulus for discussion and investigation.

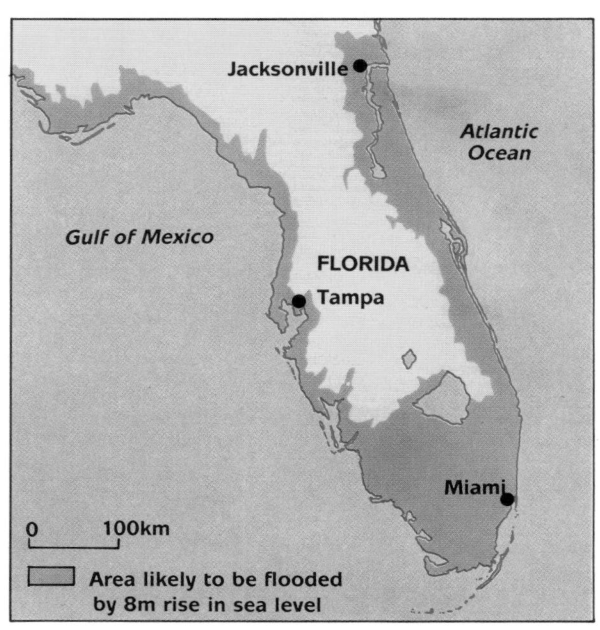


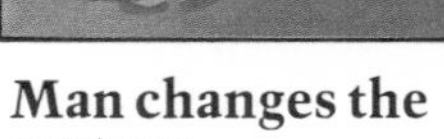

**Man changes the system**
Man's expanding agriculture and enormous consumption of fossil fuels have far-reaching consequences on the atmosphere, as they pour ever more methane and carbon dioxide into the air. Man is busy obliterating the planet's very lungs, the tropical rain forests, at the same time as increasing the workload that is required of them to extract carbon dioxide for plant growth. Temperatures are almost certain to rise; the warming of the oceans will lead to higher sea levels, and the whole system of climatic belts and habitats will undergo rapid changes. The Earth itself will certainly withstand man's assault. The question is not whether Earth will survive, but whether man will.

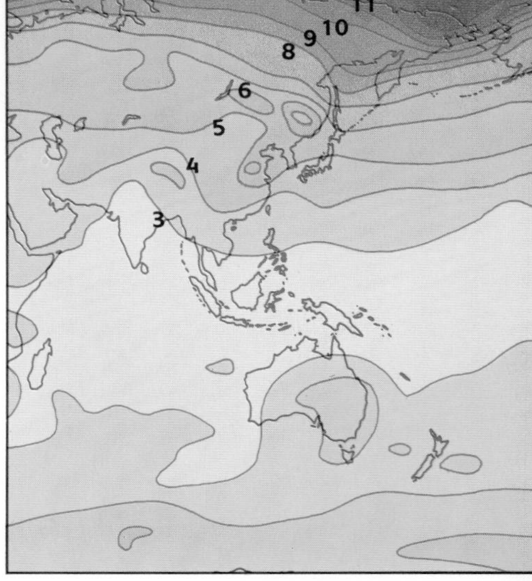


**Increase in winter surf air temperature result from a doubling of car dioxide levels in the atmosphere (°C)**

The consequences of global warming are unpredictable. A shift of climatic belts poleward could transform a temperate-latitude city like London into a near-tropical swampland.

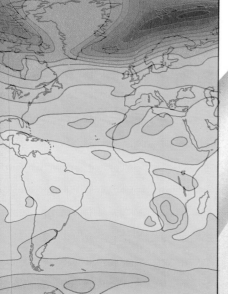

## Lovelock's model of Daisyworld

Lovelock devised a model to help explain his theory. A planet, called "Daisyworld" is covered only by light and dark daisies. The surface temperature is controlled by how much sunlight is absorbed and how much reflected back into space, a property known as "albedo." An albedo of 0 is for an object so black that no light is reflected back; an albedo of 1 is for the perfect mirror that reflects everything.

Before the daisies, the surface temperature was cool. This suited the dark daisies, as they absorbed more of the light and kept warm. However, as the dark daisies flourished, the planet's albedo was lowered, and the surface temperature began to rise. This in turn suited the light daisies, which kept cool by reflecting back most of the light. In time the planet's temperature was regulated by the balance of dark and light daisies.

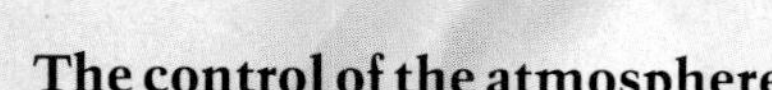

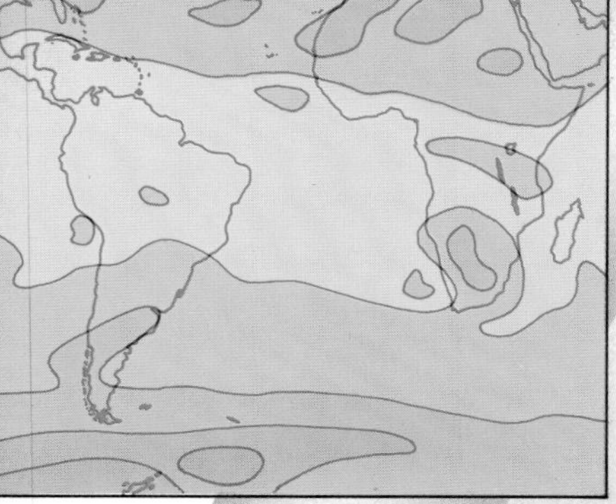

## The control of the atmosphere

The Earth's surface temperature is determined by the amount of cloud and snow cover (raising the albedo and reflecting light back into space) and the greenhouse gases in the atmosphere that trap infrared radiation reflected from the surface. The balance between these two sets of controls affects global temperatures and hence global climates.

Life directly determines the concentration of greenhouse gases. Carbon dioxide is produced by plants and animals and released by volcanoes. In turn, it is absorbed by plants in photosynthesis, while some passes into the oceans. The level of carbon dioxide can have a dramatic effect on the climate. From the analysis of bubbles of ancient air trapped in the Antarctic ice cap, we know that low levels of carbon dioxide accompanied the Ice Ages.

Although there is far less methane in the atmosphere than carbon dioxide, it is 25 times more effective at blocking infrared radiation, encouraging global warming. Methane is produced naturally, in the partial breakdown of organic matter, whether in a cow's stomach or a northern peat bog.

# GLOSSARY

**Atoll** A ring of coral islands enclosing a lagoon, generally found in the middle of the ocean, where a tower of coral has continued to grow close to the sea-surface (and the sunlight) as an underlying volcano has sunk. A drowned atoll is known as a guyot.

**Back-arc basin** New ocean *crust* formed in the wake of *subduction zones* which continue to move towards the oceanic *plate*, pulling an *island arc* away from the adjacent continent. Such back-arc basins are common around the western Pacific and in the western Mediterranean.

**Biome** Animal and plant types characteristic of a particular climatic zone, such as grassland, tropical rain forest, or desert.

**Continental shield** Mountain ranges that have been worn down over hundreds of millions of years to become vast plains, underlain by ancient crystalline rocks, formerly buried 6–12 miles (10–20km) underground.

**Convection current** In any liquid or gas, hotter material is lighter and so tends to rise, allowing colder material to take its place. This creates a circulation known as a convection current, as found within the Earth's near-liquid *mantle*.

**Crust** The outer rocks of the Earth richer in the elements silicon, aluminum, sodium and potassium (all combined with oxygen) than the underlying *mantle*. Continental crust is generally around 18 miles (28km) thick, but underneath the highest mountain ranges the crustal keel may extend to 45 miles (70km). While the crust of the continents continues to float on the *mantle* for billions of years, the thinner oceanic crust cools and sinks back into it within 200 million years.

**Evolution** Darwin's theory that.over many generations, characteristics that assist a life-form to be better adapted to its environment will tend to become enhanced. This occurs by a process of natural selection in which individuals with these characteristics tend to have more offspring.

**Fault** A breakage in the *crust* of the Earth along which the two sides have become displaced. Sudden movements along faults generate earthquakes. A "normal (or extensional) fault" involves downward motion along a sloping plane, while movement in the opposite direction defines a "reverse-fault." Where movement is sideways and the fault is almost vertical it is termed "strike-slip."

**Greenhouse effect** The warming of the Earth's atmosphere caused by the presence of molecules that are opaque to infrared light, such as carbon dioxide and methane, and which trap infrared light emitted from the ground.

**Hot-spot** A region above a rising column of hotter *mantle* material, that melts to produce large quantities of *magma*. The combination of high ground and volcanoes builds islands like Iceland and Hawaii, as well as volcanic plateaus within the continents.

**Hydrocarbon** Any of a series of chemicals composed primarily out of the elements hydrogen and carbon. Hydrocarbons develop naturally from rotting buried plant and animal remains. Gas and oil hydrocarbons rise through cracks in the crust until they may become trapped in a reservoir, bounded on its upper surface by some impermeable material, such as clay.

**Interglacial** The period lasting for tens of thousands of years between the major Ice Ages of the past two million years, when the climate is relatively mild, and ice-sheets are restricted to Greenland, Antarctica, and some high-latitude mountains. The present interglacial began around 14,000 years ago.



**Island arc** A curved line of islands that develops above a *subduction zone*, principally through the formation of a series of large volcanoes.

**Magma** Molten rock, at temperatures that may be in excess of 1,000°C, which flows or explodes from volcanoes in an eruption, or cools and hardens undergound.

**Mantle** Part of the Earth's interior lying beneath the thin outer *crust* and occupying almost 90% of the Earth by volume. Mantle rocks can be found at the Earth's surface where the mantle has been squeezed through the crust, or where fragments of rock have been broken off and erupted through volcanoes.

**Meteorite** A rock fallen through the Earth's atmosphere, from outside. Most meteorites come from the remains of a former planet, or series of small planets, that once orbited the Sun between Jupiter and Mars.

**Mineral** An inorganic , crystalline material. Almost all the solid Earth is composed of small grains of minerals, rich in the elements of silicon and oxygen, tightly set against one another.

**Organism** A living creature, with the power to sustain life through generating energy from food, and able to reproduce itself.

**Permafrost** Ground where the temperature never rises above freezing point, found in Arctic regions, and at lower latitudes on very high mountains.

**Plate tectonics** The outer layer of rocks on the Earth are cold and relatively hard, and slide as rigid "plates" about 60 miles (100km) thick, over the underlying, white-hot *mantle*. The Earth's outer shell is divided into six major plates and a number of smaller plates. Almost all movement is concentrated along the plate boundaries, zones that can be recognized from their concentration of earthquakes.

**Pull-apart** A gap that develops between strike-slip faults lying parallel with one another, leading to the creation of a near rectangular depression, often filled by a lake, beneath which accumulates a thick pile of sediments.

**Sediment** Fragments of rock, worn down by chemical processes and abrasion, in rivers, glaciers, and beaches, settle out in calm water and accumulate as layers of mud, sand, or stones. Other sediments may develop from the remains of shells, or even from minerals (like rock-salt) directly crystallized from water.

**Spreading ridge** The rifted mid-oceanic volcanic mountain range formed at the upwelling of molten rock from the *mantle* where two plates move apart.

**Subduction zone** The region where two *plates* come together and oceanic crust slides down to become, once again, part of the *mantle*.

**Terrane** A section of continental *crust*, that has become transported sideways along a plate boundary, sometimes over thousands of miles, into the midst of a mountain range.

**Tsunami** (from the Japanese "wave in the harbour") A wave, or series of waves, generated by movement of the sea floor, caused by sudden displacement on a *fault*, an underwater landslide, or a volcanic eruption. In the open ocean tsunamis are low and travel at speeds approaching that of a jet airplane, becoming slower and increasing in height as the water shallows.

# INDEX